WIND ENERGY GENERATION

Modelling and Control

WIND ENERGY GENERATION

Modelling and Control

Olimpo Anaya-Lara, *University of Strathclyde, Glasgow, UK*
Nick Jenkins, *Cardiff University, UK*
Janaka Ekanayake, *Cardiff University, UK*
Phill Cartwright, *Rolls-Royce plc, UK*
Mike Hughes, *Consultant and Imperial College London, UK*

A John Wiley and Sons, Ltd., Publication

This edition first published 2009
© 2009 John Wiley & Sons, Ltd

Registered office
John Wiley & Sons Ltd, The Atrium, Southern Gate, Chichester, West Sussex, PO19 8SQ, United Kingdom

For details of our global editorial offices, for customer services and for information about how to apply for permission to reuse the copyright material in this book please see our website at www.wiley.com.

Library of Congress Cataloguing-in-Publication Data

Wind energy generation : modelling and control / Olimpo Anaya-Lara ... [et al.].
     p. cm.
  Includes index.
  ISBN 978-0-470-71433-1 (cloth)
  1. Wind power. 2. Wind turbines. 3. Synchronous generators. I. Anaya-Lara,
Olimpo.
  TJ820.W56955 2009
  621.31'2136–dc22

                                         2009012004

ISBN: 978-0-470-71433-1 (HB)

A catalogue record for this book is available from the British Library.

Typeset in 11/13.5pt Times-Roman by Laserwords Private Limited, Chennai, India.
Printed and bound in Great Britain by CPI, Anthony Rowe, Chippenham, Wiltshire

Contents

About the Authors

Olimpo Anaya-Lara is a Lecturer in the Institute for Energy and Environment at the University of Strathclyde, UK. Over the course of his career, he has successfully undertaken research on power electronic equipment, control systems development, and stability and control of power systems with increased wind energy penetration. He was a member of the International Energy Agency Annexes XXI *Dynamic models of wind farms for power system studies* and XXIII *Offshore wind energy technology development*. He is currently a Member of the IEEE and IET, and has published 2 technical books, as well as over 80 papers in international journals and conference proceedings.

Nick Jenkins was at the University of Manchester (UMIST) from 1992 to 2008. In 2008 he moved to Cardiff University where he is now the Professor of Renewable Energy. His career includes 14 years of industrial experience, 5 of which were spent in developing countries. His final position before joining the university was as a Projects Director for the Wind Energy Group, a manufacturer of large wind turbines. He is a Fellow of the IET, IEEE and Royal Academy of Engineering. In 2009 and 2010 he was the Shimizu visiting professor at Stanford University.

Janaka Ekanayake joined Cardiff University as a Senior Lecturer in June 2008 from the University of Manchester where he was a Research Fellow. Since 1992 he has been attached to the University of Peradeniya, Sri Lanka and was promoted to a Professor in Electrical and Electronic Engineering in 2003. He is a Senior Member of the IEEE and a Member of IET. His main research interests include power electronic applications for power systems, renewable energy generation and its integration. He has published more than 25 papers in refereed journals and has also coauthored a book.

Phill Cartwright has 20 years of industrial experience in the research, analyses, design and implementation of flexible power systems architectures and projects with ABB, ALSTOM and AREVA in Brazil, China, Europe, India and the USA. He is currently the Head of the global Electrical & Automation Systems business for Rolls-Royce Group Plc, providing integrated power systems products and technology for Civil Aerospace, Defence Aerospace, Marine Systems, New Nuclear and emerging Tidal Generation markets and developments. He is a visiting professor in Power Systems at The University of Strathclyde, UK.

Mike Hughes graduated from the University of Liverpool in 1961 with first class honours in electrical engineering. His initial career in the power industry was with the Associated Electrical Industries and The Nuclear Power Group, working on network analysis and control scheme design. From 1971 to 1999, he was with the University of Manchester Institute of Science and Technology teaching and researching in the areas of power system dynamics and control. He is currently a part-time Research Fellow with Imperial College, London and a consultant in power plant control and wind generation systems.

Preface

The stimulus for this book is the rapid expansion worldwide of wind energy systems and the implications that this has for power system operation and control. Rapidly evolving wind turbine technology and the widespread use of advanced power electronic converters call for more detailed and accurate modelling of the various components involved in wind energy systems and their controllers. As wind turbine technology differs significantly from that employed by conventional generating plants based on synchronous generators, the dynamic characteristics of the electrical power network may be drastically changed and hence the requirements for network control and operation may also be different. In addition, new Grid Code regulations for connection of large wind farms now impose the requirement that wind farms should be able to contribute to network support and operation as do conventional generation plants based on synchronous generators. To address these challenges good knowledge of wind generation dynamic models, control capabilities and interaction with the power system becomes critical.

The book aims to provide a basic understanding of modelling of wind generation systems, including both the mechanical and electrical systems, and to examine the control philosophies and schemes that enable the reliable, secure and cost-effective operation of these generation systems. The book is intended for later year undergraduate and post-graduate students interested in understanding the modelling and control of large wind turbine generators, as well as practising engineers and those responsible for grid integration. It starts with a review of the principles of operation, modelling and control of the common wind generation systems used and then moves on to discuss grid compatibility and the influence of wind turbines on power system operation and stability.

Chapter 1 provides an overview of the current status of wind energy around the world and introduces the most commonly used wind turbine configurations. Typical converter topologies and pulse-width modulation control techniques used in wind generation systems are presented in Chapter 2. Chapter 3 introduces fundamental knowledge for the mathematical modelling of synchronous machines and their representation for transient stability studies. Chapters 4 to

6 present the mathematical modelling of fixed-speed and variable-speed wind turbines, introducing typical control methodologies. Dynamic performance under small and large network disturbances is illustrated through various case studies. Different representations of shaft and blade dynamics are explained in Chapter 7 to illustrate how structural dynamics affect the performance of the wind turbine during electrical transients. The interaction between bulk wind farm generation and conventional generation and its influence on network dynamic characteristics are explained in Chapter 8. Time response simulation and eigenvalue analysis are used to establish basic transient and dynamic stability characteristics. This then leads into Chapter 9 where more advanced control strategies for variable-speed wind turbines are addressed such as the inclusion of a power system stabiliser. Enabling technologies for wind farm integration are discussed in Chapter 10 and finally Chapter 11 presents different ways in which the wind turbine can be controlled for system contingencies.

The text presented in this book draws together material on modelling and control of wind turbines from many sources, e.g. graduate courses that the authors have taught over many years at universities in the UK, USA, Sri Lanka and Mexico, a large number of technical papers published by the IEEE and IET, and research programmes with which they have been closely associated such as the EPSRC-funded SUPERGEN Future Network Technologies and the DECC-funded UK SEDG. Through these programmes the authors have had the chance to interact closely with industrial partners (utilities, power electronic equipment manufacturers and wind farm developers) and get useful points of view on the needs and priorities of the wind energy sector concerning wind turbine generator dynamic modelling and control. The authors would like to thank Prof. Jim McDonald and Prof. Goran Strbac, co-directors of the UK SEDG. Thanks are also given to Dr. Nolan Caliao and Mr. Piyadanai Pachanapan who assisted in the preparation of drawings, to Dr. Gustavo Quinonez-Varela who provided input into the operation of fixed-speed wind turbines, and to Ms Rose King who provided useful material for Chapter 10. Special thanks go to Dr. Ramtharan Gnanasambandapillai who gave permission to include material from his PhD thesis in Chapter 7.

Olimpo Anaya-Lara
Nick Jenkins
Janaka Ekanayake
Phill Cartwright
Mike Hughes
2009

Acronyms and Symbols

AC	Alternating current
AVR	Automatic voltage regulator
CB-PWM	Carrier-based PWM
DC	Direct current
DFIG	Doubly fed induction generator
emf	Electromotive force
FC	Fixed capacitor
FMAC	Flux magnitude and angle controller
FRC	Fully rated converter
FRC-SG	Fully rated converter wind turbine using synchronous generator
FRT	Fault ride-through
FSIG	Fixed-speed induction generator
GSC	Generator-side converter
HVAC	High-voltage alternating current
HVDC	High-voltage direct current
IGBT	Insulated-gate bipolar transistor
LCC-HVDC	Line-commutated converter HVDC
NRS-PWM	Non-regular sampled PWM
NSC	Network-side converter
PAM	Pulse-amplitude modulation
PI	Proportional–integral controller
PLL	Phase-locked loop
PM	Permanent magnet
PoC	Point of connection
PPC	Power production control
PSS	Power system stabilizer
pu	Per unit
PWM	Pulse-width modulation
RMS	Root mean square
RPM	Revolutions per minute

RS-PWM	Regular sampled PWM
SFO	Stator flux oriented
SFO-PWM	Switching frequency optimal PWM
SHEM-PWM	Selective harmonic elimination PWM
STATCOM	Static compensator
SVC	Static var compensator
SV-PWM	Space vector PWM
TCR	Thyristor-controlled reactor
TSC	Thyristor-switched capacitor
VSC	Voltage source converter
VSC–HVDC	Voltage source converter HVDC
P_{air}	Power in the airflow
ρ	Air density
A	Swept area of rotor, m^2
ν	Upwind free wind speed, ms^{-1}
C_p	Power coefficient
$P_{wind\ turbine}$	Power transferred to the wind turbine rotor
λ	Tip-speed ratio
ω	Rotational speed of rotor
R	Radius to tip of rotor
V_m	Mean annual site wind speed
V_{DC}	Direct voltage
Over$^-$	Per unit quantity
b	Base quantity
ϕ_s	Stator magnetic field
ϕ_r	Rotor magnetic field
i_{ds}, i_{qs}	Stator currents in d and q axis
v_{ds}, v_{qs}	Stator voltages in d and q axis
ψ_{ds}, ψ_{qs}	Stator flux linkage in d and q axis
T_e	Electromagnetic torque
T_m	Mechanical torque
P_e	Electrical power
P_m	Mechanical power
Q	Reactive power
ω_b	Base synchronous speed
ω_s	Synchronous speed
ω_r	Rotor speed
J	Inertia constant
H	Per unit inertia constant
K	Shaft stiffness

f	System frequency
C	Capacitance

Synchronous Generator

i_f	Field current
i_{kd}, i_{kq1}, i_{kq2}	Damper winding d and q axis currents
L_{lkd}, L_{lkq}	Leakage inductance of damper windings in d and q axis
L_{md}, L_{mq}	Mutual inductance in d and q axis
L_{lf}	Leakage inductance of the field coil
L_{ls}	Leakage inductance of the stator coil
r_s	Stator resistance
r_f	Field winding resistance
r_{kd}, r_{kq1}, r_{kq2}	Resistance of damper d and q axis coils
v_{fd}	Field voltage
v_{kd}, v_{kq1}, v_{kq2}	Damper winding voltages in d and q axis
ψ_f	Field flux linkage
ψ_{kd}, ψ_{kq1}, ψ_{kq2}	Damper winding flux linkage in d and q axis
δ_r	Rotor angle
C_s	Synchronizing power coefficient
C_d	Damping power coefficient

Induction Generator

i_{dr}, i_{qr}	Rotor currents in d and q axis
v_{dr}, v_{qr}	Rotor voltages in d and q axis
ψ_{dr}, ψ_{qr}	Rotor flux linkage in d and q axis
e_d, e_q	Voltage behind a transient reactance in d and q axis
L_m	Mutual inductance between stator and rotor windings
X_m	Magnetizing reactance
L_r, L_s	Rotor and stator self-inductance
X_r, X_s	Rotor and stator reactance
L_{lr}	Rotor leakage inductance
L_{ls}	Stator leakage inductance
r_r	Rotor resistance
r_s	Stator resistance
s	Slip of an induction generator
p	Number of poles

1

Electricity Generation from Wind Energy

There is now general acceptance that the burning of fossil fuels is having a significant influence on the global climate. Effective mitigation of climate change will require deep reductions in greenhouse gas emissions, with UK estimates of a 60–80% cut being necessary by 2050 (Stern Review, UK HM Treasury, 2006). The electricity system is viewed as being easier to transfer to low-carbon energy sources than more challenging sectors of the economy such as surface and air transport and domestic heating. Hence the use of cost-effective and reliable low-carbon electricity generation sources, in addition to demand-side measures, is becoming an important objective of energy policy in many countries (EWEA, 2006; AWEA, 2007).

Over the past few years, wind energy has shown the fastest rate of growth of any form of electricity generation with its development stimulated by concerns of national policy makers over climate change, energy diversity and security of supply.

Figure 1.1 shows the global cumulative wind power capacity worldwide (GWEC, 2006). In this figure, the 'Reference' scenario is based on the projection in the 2004 World Energy Outlook report from the International Energy Agency (IEA). This projects the growth of all renewables including wind power, up to 2030. The 'Moderate' scenario takes into account all policy measures to support renewable energy either under way or planned worldwide. The 'Advanced' scenario makes the assumption that all policy options are in favour of wind power, and the political will is there to carry them out.

Wind Energy Generation: Modelling and Control   Olimpo Anaya-Lara, Nick Jenkins, Janaka Ekanayake, Phill Cartwright and Mike Hughes
© 2009 John Wiley & Sons, Ltd

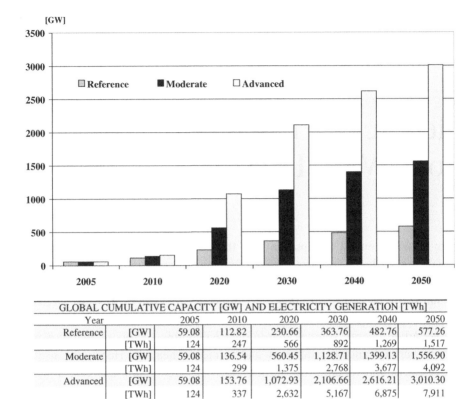

GLOBAL CUMULATIVE CAPACITY [GW] AND ELECTRICITY GENERATION [TWh]							
Year		2005	2010	2020	2030	2040	2050
Reference	[GW]	59.08	112.82	230.66	363.76	482.76	577.26
	[TWh]	124	247	566	892	1,269	1,517
Moderate	[GW]	59.08	136.54	560.45	1,128.71	1,399.13	1,556.90
	[TWh]	124	299	1,375	2,768	3,677	4,092
Advanced	[GW]	59.08	153.76	1,072.93	2,106.66	2,616.21	3,010.30
	[TWh]	124	337	2,632	5,167	6,875	7,911

Figure 1.1 Global cumulative wind power capacity (GWEC, 2006)

1.1 Wind Farms

Numerous wind farm projects are being constructed around the globe with both offshore and onshore developments in Europe and primarily large onshore developments in North America. Usually, sites are preselected based on general information of wind speeds provided by a wind atlas, which is then validated with local measurements. The local wind resource is monitored for 1 year, or more, before the project is approved and the wind turbines installed.

Onshore turbine installations are frequently in upland terrain to exploit the higher wind speeds. However, wind farm permitting and siting onshore can be difficult as high wind-speed sites are often of high visual amenity value and environmentally sensitive.

Offshore development, particularly of larger wind farms, generally takes place more than 5 km from land to reduce environmental impact. The advantages of offshore wind farms include reduced visual intrusion and acoustic noise impact and also lower wind turbulence with higher average wind speeds.

Table 1.1   Wind turbine applications (Elliot, 2002)

Small (≤10 kW)	Intermediate (10–500 kW)	Large (500 kW–5 MW)
• Homes (grid-connected) • Farms • Autonomous remote applications (e.g. battery charging, water pumping, telecom sites)	• Village power • Hybrid systems • Distributed power	• Wind power plants • Distributed power • Onshore and offshore wind generation

The obvious disadvantages are the higher costs of constructing and operating wind turbines offshore, and the longer power cables that must be used to connect the wind farm to the terrestrial power grid.

In general, the areas of good wind energy resource are found far from population centres and new transmission circuits are needed to connect the wind farms into the main power grid. For example, it is estimated that in Germany, approximately 1400 km of additional high-voltage and extra-high-voltage lines will be required over the next 10 years to connect new wind farms (Deutsche Energie-Agentur GmbH, 2005).

Smaller wind turbines may also be used for rural electrification with applications including village power systems and stand-alone wind systems for hospitals, homes and community centres (Elliot, 2002).

Table 1.1 illustrates typical wind turbine ratings according to their application.

1.2   Wind Energy-generating Systems

Wind energy technology has evolved rapidly over the last three decades (Figure 1.2) with increasing rotor diameters and the use of sophisticated power electronics to allow operation at variable rotor speed.

1.2.1   Wind Turbines

Wind turbines produce electricity by using the power of the wind to drive an electrical generator. Wind passes over the blades, generating lift and exerting a turning force. The rotating blades turn a shaft inside the nacelle, which goes into a gearbox. The gearbox increases the rotational speed to that which is appropriate for the generator, which uses magnetic fields to convert the rotational energy into electrical energy. The power output goes to a transformer, which converts the electricity from the generator at around 700 V to the appropriate voltage for the power collection system, typically 33 kV.

A wind turbine extracts kinetic energy from the swept area of the blades (Figure 1.3). The power in the airflow is given by (Manwell *et al.*, 2002;

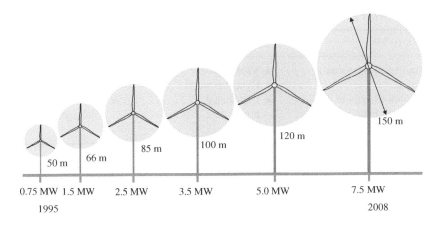

Figure 1.2   Evolution of wind turbine dimensions

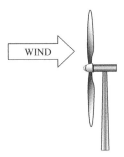

Figure 1.3   Horizontal axis wind turbine

Burton *et al.*, 2001):

$$P_{\text{air}} = \frac{1}{2}\rho A v^3 \tag{1.1}$$

where

ρ = air density (approximately $1.225\,\text{kg m}^{-3}$)
A = swept area of rotor, m^2
v = upwind free wind speed, m s^{-1}.

Although Eq. (1.1) gives the power available in the wind the power transferred to the wind turbine rotor is reduced by the power coefficient, C_p:

$$C_p = \frac{P_{\text{wind turbine}}}{P_{\text{air}}} \tag{1.2}$$

$$P_{\text{wind turbine}} = C_p P_{\text{air}} = C_p \times \frac{1}{2}\rho A v^3 \tag{1.3}$$

A maximum value of C_p is defined by the Betz limit, which states that a turbine can never extract more than 59.3% of the power from an air stream. In reality, wind turbine rotors have maximum C_p values in the range 25–45%.

It is also conventional to define a tip-speed ratio, λ, as

$$\lambda = \frac{\omega R}{v} \qquad (1.4)$$

where

$\omega =$ rotational speed of rotor

$R =$ radius to tip of rotor

$v =$ upwind free wind speed, $m\,s^{-1}$.

The tip-speed ratio, λ, and the power coefficient, C_p, are dimensionless and so can be used to describe the performance of any size of wind turbine rotor. Figure 1.4 shows that the maximum power coefficient is only achieved at a single tip-speed ratio and for a fixed rotational speed of the wind turbine this only occurs at a single wind speed. Hence, one argument for operating a wind turbine at variable rotational speed is that it is possible to operate at maximum C_p over a range of wind speeds.

The power output of a wind turbine at various wind speeds is conventionally described by its power curve. The power curve gives the steady-state electrical power output as a function of the wind speed at the hub height and is generally

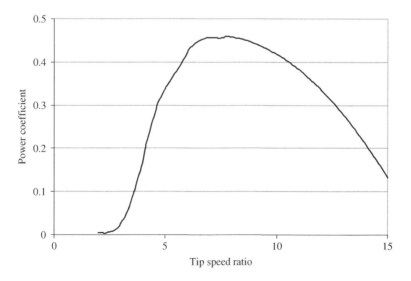

Figure 1.4   Illustration of power coefficient/tip-speed ratio curve, C_p/λ

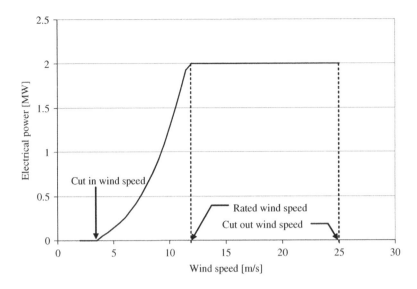

Figure 1.5   Power curve for a 2 MW wind turbine

measured using 10 min average data. An example of a power curve is given in Figure 1.5.

The power curve has three key points on the velocity scale:

- Cut-in wind speed – the minimum wind speed at which the machine will deliver useful power.
- Rated wind speed – the wind speed at which rated power is obtained (rated power is generally the maximum power output of the electrical generator).
- Cut-out wind speed – the maximum wind speed at which the turbine is allowed to deliver power (usually limited by engineering loads and safety constraints).

Below the cut-in speed, of about $5 \, \mathrm{m \, s^{-1}}$, the wind turbine remains shut down as the speed of the wind is too low for useful energy production. Then, once in operation, the power output increases following a broadly cubic relationship with wind speed (although modified by the variation in C_p) until rated wind speed is reached. Above rated wind speed the aerodynamic rotor is arranged to limit the mechanical power extracted from the wind and so reduce the mechanical loads on the drive train. Then at very high wind speeds the turbine is shut down.

The choice of cut-in, rated and cut-out wind speed is made by the wind turbine designer who, for typical wind conditions, will try to balance obtaining maximum energy extraction with controlling the mechanical loads (and hence the capital cost) of the turbine. For a mean annual site wind speed V_m of $8\,m\,s^{-1}$ typical values will be approximately (Fox *et al.*, 2007):

- cut-in wind speed: $5\,m\,s^{-1}$, $0.6\,V_m$
- rated wind speed: $12-14\,m\,s^{-1}$, $1.5-1.75\,V_m$
- cut-out wind speed: $25\,m\,s^{-1}$, $3V_m$.

Power curves for existing machines can normally be obtained from the turbine manufacturer. They are found by field measurements, where an anemometer is placed on a mast reasonably close to the wind turbine, not on the turbine itself or too close to it, since the turbine may create turbulence and make wind speed measurements unreliable.

1.2.2   Wind Turbine Architectures

There are a large number of choices of architecture available to the designer of a wind turbine and, over the years, most of these have been explored (Ackermann, 2005; Heier, 2006). However, commercial designs for electricity generation have now converged to horizontal axis, three-bladed, upwind turbines. The largest machines tend to operate at variable speed whereas smaller, simpler turbines are of fixed speed.

Modern electricity-generating wind turbines now use three-bladed upwind rotors, although two-bladed, and even one-bladed, rotors were used in earlier commercial turbines. Reducing the number of blades means that the rotor has to operate at a higher rotational speed in order to extract the wind energy passing through the rotor disk. Although a high rotor speed is attractive in that it reduces the gearbox ratio required, a high blade tip speed leads to increased aerodynamic noise and increased blade drag losses. Most importantly, three-bladed rotors are visually more pleasing than other designs and so these are now always used on large electricity-generating turbines.

1.2.2.1   Fixed-speed Wind Turbines

Fixed-speed wind turbines are electrically fairly simple devices consisting of an aerodynamic rotor driving a low-speed shaft, a gearbox, a high-speed shaft and an induction (sometimes known as asynchronous) generator. From the electrical system viewpoint they are perhaps best considered as large fan drives with torque applied to the low-speed shaft from the wind flow.

Figure 1.6 illustrates the configuration of a fixed-speed wind turbine (Holdsworth *et al.*, 2003; Akhmatov, 2007). It consists of a squirrel-cage

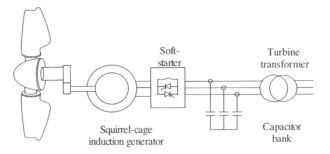

Figure 1.6   Schematic of a fixed-speed wind turbine

induction generator coupled to the power system through a turbine transformer. The generator operating slip changes slightly as the operating power
level changes and the rotational speed is therefore not entirely constant.
However, because the operating slip variation is generally less than 1%, this
type of wind generation is normally referred to as fixed speed.

Squirrel-cage induction machines consume reactive power and so it is conventional to provide power factor correction capacitors at each wind turbine.
The function of the soft-starter unit is to build up the magnetic flux slowly
and so minimize transient currents during energization of the generator. Also,
by applying the network voltage slowly to the generator, once energized, it
brings the drive train slowly to its operating rotational speed.

1.2.2.2   Variable-speed Wind Turbines

As the size of wind turbines has become larger, the technology has switched
from fixed speed to variable speed. The drivers behind these developments
are mainly the ability to comply with Grid Code connection requirements
and the reduction in mechanical loads achieved with variable-speed operation.
Currently the most common variable-speed wind turbine configurations are as
follows:

- doubly fed induction generator (DFIG) wind turbine
- fully rated converter (FRC) wind turbine based on a synchronous or
  induction generator.

Doubly Fed Induction Generator (DFIG) Wind Turbine
A typical configuration of a DFIG wind turbine is shown schematically
in Figure 1.7. It uses a wound-rotor induction generator with slip rings to
take current into or out of the rotor winding and variable-speed operation is

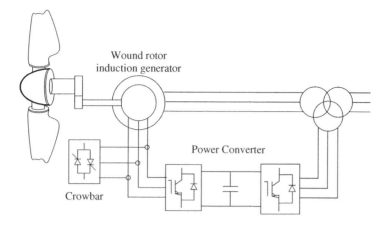

Figure 1.7   Typical configuration of a DFIG wind turbine

obtained by injecting a controllable voltage into the rotor at slip frequency (Müller *et al.*, 2002; Holdsworth *et al.*, 2003). The rotor winding is fed through a variable-frequency power converter, typically based on two AC/DC IGBT-based voltage source converters (VSCs), linked by a DC bus. The power converter decouples the network electrical frequency from the rotor mechanical frequency, enabling variable-speed operation of the wind turbine. The generator and converters are protected by voltage limits and an over-current 'crowbar'.

A DFIG system can deliver power to the grid through the stator and rotor, while the rotor can also absorb power. This depends on the rotational speed of the generator. If the generator operates above synchronous speed, power will be delivered from the rotor through the converters to the network, and if the generator operates below synchronous speed, then the rotor will absorb power from the network through the converters.

Fully Rated Converter (FRC) Wind Turbine

The typical configuration of a fully rated converter wind turbine is shown in Figure 1.8. This type of turbine may or may not include a gearbox and a wide range of electrical generator types can be employed, for example, induction, wound-rotor synchronous or permanent magnet synchronous. As all of the power from the turbine goes through the power converters, the dynamic operation of the electrical generator is effectively isolated from the power grid (Akhmatov *et al.*, 2003; Heier, 2006). The electrical frequency of the generator may vary as the wind speed changes, while the grid frequency remains unchanged, thus allowing variable-speed operation of the wind turbine.

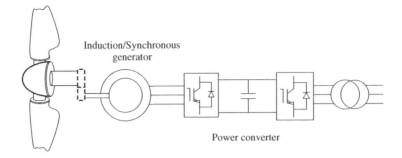

Figure 1.8 Typical configuration of a fully rated converter-connected wind turbine

The power converters can be arranged in various ways. Whereas the generator-side converter (GSC) can be a diode rectifier or a PWM voltage source converter (VSC), the network-side converter (NSC) is typically a PWM VSC. The strategy to control the operation of the generator and the power flows to the network depends very much on the type of power converter arrangement employed. The network-side converter can be arranged to maintain the DC bus voltage constant with torque applied to the generator controlled from the generator-side converter. Alternatively, the control philosophy can be reversed. Active power is transmitted through the converters with very little energy stored in the DC link capacitor. Hence the torque applied to the generator can be controlled by the network-side converter. Each converter is able to generate or absorb reactive power independently.

1.3 Wind Generators Compared with Conventional Power Plant

There are significant differences between wind power and conventional synchronous central generation (Slootweg, 2003):

- Wind turbines employ different, often converter-based, generating systems compared with those used in conventional power plants.
- The prime mover of wind turbines, the wind, is not controllable and fluctuates stochastically.
- The typical size of individual wind turbines is much smaller than that of a conventional utility synchronous generator.

Due to these differences, wind generation interacts differently with the network and wind generation may have both local and system-wide impacts on the operation of the power system. Local impacts occur in the electrical vicinity of a wind turbine or wind farm, and can be attributed to a specific turbine or farm.

System-wide impacts, on the other hand, affect the behaviour of the power system as a whole. They are an inherent consequence of the utilization of wind power and cannot be attributed to individual turbines or farms (UCTE, 2004).

1.3.1 Local Impacts

Locally, wind power has an impact on the following aspects of the power system:

- circuit power flows and busbar voltages
- protection schemes, fault currents, and switchgear rating
- power quality
  - harmonic voltage distortion
  - voltage flicker.

The first two topics are always investigated when connecting any new generator and are not specific to wind power. Harmonic voltage distortion is of particular interest when power electronic converters are employed to interface wind generation units to the network whereas voltage flicker is more significant for large, fixed-speed wind turbines on weak distribution circuits.

1.3.1.1 Circuit Power Flows and Busbar Voltages

The way in which wind turbines affect locally the circuit active and reactive power flows and busbar voltages depends on whether fixed-speed or variable-speed turbines are used. The operating condition of a squirrel-cage induction generator, used in fixed-speed turbines, is dictated by the mechanical input power and the voltage at the generator terminals. This type of generator cannot control busbar voltages by itself controlling the reactive power exchange with the network. Additional reactive power compensation equipment, often fixed shunt-connected capacitors, is normally fitted. Variable-speed turbines have, in principle, the capability of varying the reactive power that they exchange with the grid to affect their terminal voltage. In practice, this capability depends to a large extent on the rating and the controllers of the power electronic converters.

1.3.1.2 Protection Schemes, Fault Currents and Switchgear Rating

The contribution of wind turbines to network fault current also depends on the generator technology employed. Fixed-speed turbines, in common with all directly connected spinning plant, contribute to network fault currents.

However, as they use induction generators, they contribute only sub-transient fault current (lasting less than, say, 200 ms) to balanced three-phase faults but can supply sustained fault current to unbalanced faults. They rely on sequential tripping (over/under-voltage, over/under-frequency and loss of mains) protection schemes to detect when conventional over-current protection has isolated a faulty section of the network to which they are connected.

Variable-speed DFIG wind turbines also contribute to network fault currents with the control system of the power electronic converters detecting the fault very quickly. Due to the sensitivity of the power electronics to over-currents, this type of wind turbine may be quickly disconnected from the network and the crowbar activated to short-circuit the rotor windings of the wound-rotor induction generator, unless special precautions are taken to ensure Grid Code compliance.

Fully rated converter-connected wind turbines generally do not contribute significantly to network fault current because the network-side converter is not sized to supply sustained over-currents. Again, this wind turbine type may also disconnect quickly in the case of a fault, if the Grid Codes do not require a Fault Ride Through capability.

The behaviour of power converter-connected wind turbines during network faults depends on the design of the power converters and the settings of their control systems. There are as yet no agreed international standards for either the fault contribution performance required of converter-connected generators or how such generators should be represented in transient stability or fault calculator simulation programs. A conservative design approach is to assume that such generators do contribute fault current when rating switchgear and other plant, but not to rely on such fault currents for protection operation.

1.3.1.3 Power Quality

Two local effects of wind power on power quality may be considered, voltage harmonic distortion and flicker. Harmonic distortion is mainly associated with variable-speed wind turbines because these contain power electronic converters, which are an important source of high-frequency harmonic currents. It is increasingly of concern in large offshore wind farms, where the very extensive cable networks can lead to harmonic resonances and high harmonic currents caused by existing harmonic voltages already present on the power system or by the wind turbine converters.

In fixed-speed wind turbines, wind fluctuations are directly translated into output power fluctuations because there is no energy buffer between the mechanical input and the electrical output. Depending on the strength of the grid connection, the resulting power fluctuations can result in grid voltage fluctuations, which can cause unwanted and annoying fluctuations in electric light bulb brightness. This problem is referred to as 'flicker'. In general, flicker problems do not occur with variable-speed turbines, because in these turbines wind speed fluctuations are not directly translated into output power fluctuations. The stored energy of the spinning mass of the rotor acts as an energy buffer.

1.3.2   System-wide Impacts

In addition to the local impacts, wind power also has a number of system-wide impacts as it affects the following (Slootweg, 2003; UCTE, 2004):

- power system dynamics and stability
- reactive power and voltage support
- frequency support.

1.3.2.1   Power System Dynamics and Stability

Squirrel-cage induction generators used in fixed-speed turbines can cause local voltage collapse after rotor speed runaway. During a fault (and consequent network voltage depression), they accelerate due to the imbalance between the mechanical power from the wind and the electrical power that can be supplied to the grid. When the fault is cleared, they absorb reactive power, depressing the network voltage. If the voltage does not recover quickly enough, the wind turbines continue to accelerate and to consume large amounts of reactive power. This eventually leads to voltage and rotor speed instability. In contrast to synchronous generators, the exciters of which increase reactive power output during low network voltages and thus support voltage recovery after a fault, squirrel-cage induction generators tend to impede voltage recovery.

With variable-speed wind turbines, the sensitivity of the power electronics to over-currents caused by network voltage depressions can have serious consequences for the stability of the power system. If the penetration level of variable-speed wind turbines in the system is high and they disconnect at relatively small voltage reductions, a voltage drop over a wide geographic area can lead to a large generation deficit. Such a voltage drop could be caused, for instance, by a fault in the transmission grid. To prevent this, grid companies and transmission system operators require that wind turbines have a Fault Ride Through capability and are able to withstand voltage drops of certain magnitudes and durations without tripping. This prevents the disconnection of a large amount of wind power in the event of a remote network fault.

1.3.2.2 Reactive Power and Voltage Support

The voltage on a transmission network is determined mainly by the inter-action of reactive power flows with the reactive inductance of the network. Fixed-speed induction generators absorb reactive power to maintain their magnetic field and have no direct control over their reactive power flow. Therefore, in the case of fixed-speed induction generators, the only way to support the voltage of the network is to reduce the reactive power drawn from the network by the use of shunt compensators.

Variable-speed wind turbines have the capability of reactive power control and may be able to support the voltage of the network to which they are connected. However, individual control of wind turbines may not be able to control the voltage at the point of connection, especially because the wind farm network is predominantly capacitive (a cable network).

On many occasions, the reactive power and voltage control at the point of connection of the wind farm is achieved by using reactive power compensation equipment such as static var compensators (SVCs) or static synchronous compensators (STATCOMs).

1.3.2.3 Frequency Support

To provide frequency support from a generation unit, the generator power must increase or decrease as the system frequency changes. Hence, in order to respond to low network frequency, it is necessary to de-load the wind turbine leaving a margin for power increase. A fixed-speed wind turbine can be de-loaded if the pitch angle is controlled such that a fraction of the power that could be extracted from wind will be 'spilled'. A variable-speed wind turbine can be de-loaded by operating it away from the maximum power extraction curve, thus leaving a margin for frequency control.

1.4 Grid Code Regulations for the Integration of Wind Generation

Grid connection codes define the requirements for the connection of generation and loads to an electrical network which ensure efficient, safe and economic operation of the transmission and/or distribution systems. Grid Codes specify the mandatory minimum technical requirements that a power plant should fulfil and additional support that may be called on to maintain the second-by-second power balance and maintain the required level of quality and security of the system. The additional services that a power plant should provide are normally

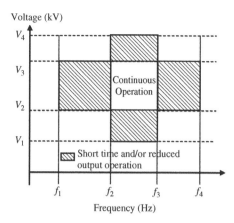

Figure 1.9 Typical shape of continuous and reduced output regions (after Great Britain and Ireland Grid Codes; ESB National Grid, 2008; National Grid, 2008)

agreed between the transmission system operator and the power plant operator through market mechanisms.

The connection codes normally focus on the point of connection between the public electricity system and the new generation. This is very important for wind farm connections, as the Grid Codes demand requirements at the point of connection of the wind farm not at the individual wind turbine generator terminals. The grid connection requirements differ from country to country and may differ from region to region. They have many common features but some of the requirements are subtly different, reflecting the characteristics of the individual grids.

As a mandatory requirement, the levels and time period of the output power of a generating plant should be maintained within the specified values of grid frequency and grid voltage as specified in Grid Codes. Typically, this requirement is defined as shown in Figure 1.9, where the values of voltage, V_1 to V_4, and frequency, f_1 to f_4, differ from country to country.

Grid Codes also specify the steady-state operational region of a power plant in terms of active and reactive power requirements. The definition of the operational region differs from country to country. For example, Figure 1.10 shows the operational regions as specified in the Great Britain and Ireland Grid Codes.

Almost all Grid Codes now impose the requirement that wind farms should be able to provide primary frequency response. The capability profile typically specifies the minimum required level of response, the frequency deviation at which it should be activated and time to respond.

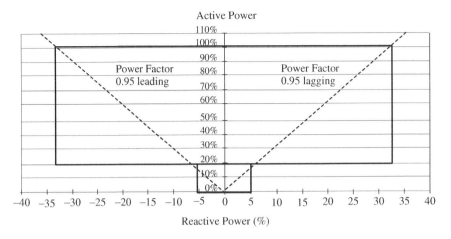

Figure 1.10 Typical steady-state operating region (after Great Britain and Ireland Grid Codes; ESB National Grid, 2008; National Grid, 2008)

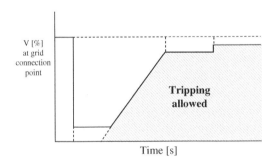

Figure 1.11 Typical shape of Fault Ride Through capability plot (after Great Britain and Ireland Grid Codes; ESB National Grid, 2008; National Grid, 2008)

Traditionally, wind turbine generators were tripped off once the voltage at their terminals reduced to a specified level. However, with the penetration of wind generation increasing, Grid Codes now generally demand Fault Ride Through capability for wind turbines connected to transmission networks. Figure 1.11 shows a plot illustrating the general shape of voltage tolerance that most grid operators demand. When reduced system voltage occurs following a network fault, generator tripping is only permitted when the voltage is sufficiently low and for a time that puts it in the shaded area shown in Figure 1.11. Grid Codes are under continual review and, as the level of wind power increases, are likely to become mode demanding.

References

Ackermann, T. (ed.) (2005) *Wind Power in Power Systems*, John Wiley & Sons, Ltd, Chichester, ISBN 10: 0470855088.

Akhmatov, V. (2007) *Induction Generators for Wind Power*, Multi-Science Publishing, Brentwood, ISBN 10: 0906522404.

Akhmatov, V., Nielsen, A. F., Pedersen, J. K. and Nymann, O. (2003) Variable-speed wind turbines with multi-pole synchronous permanent magnet generators. Part 1: modelling in dynamic simulation tools, *Wind Engineering*, **27**, 531–548.

AWEA, American Wind Energy Association (2007) *Wind Web Tutorial*, www.awea.org/faq/index.html; last accessed 18 March 2009.

Burton, T., Sharpe, D., Jenkins, N. and Bossanyi, E. (2001) *Wind Energy Handbook*, John Wiley & Sons, Ltd, Chichester, ISBN 10: 0471489972.

Deutsche Energie-Agentur GmbH (2005) *Planning of the Grid Integration of Wind Energy in Germany Onshore and Offshore up to the Year 2020 (Summary of the Essential Results of the Dena Grid Study)*. Summary available online at: www.eon-netz.com/Ressources/downloads/dena-Summary-Consortium-English.pdf; last accessed 18 March 2009

ESB National Grid (2008), *EirGrid – Grid Code*, www.eirgrid.com/EirgridPortal/default.aspx?tabid=Grid%20Code; last accessed 18 March 2009.

Elliot, D. (2002) Assessing the world's wind resources, *IEEE Power Engineering Review*, **22** (9), 4–9.

EWEA, European Wind Energy Association (2006) *Large Scale Integration of Wind Energy in the European Power Supply: Analysis, Issues and Recommendations A Report by EWEA*. EWEA, Brussels.

Fox, B., Flynn, D., Bryans, L., Jenkins, N., Milborrow, D., O'Malley, M., Watson, R. and Anaya-Lara, O. (2007) *Wind Power Integration: Connection and System Operational Aspects*, IET Power and Energy Series 50, Institution of Engineering and Technology, Stevenage, ISBN 10: 0863414494.

GWEC, Global Wind Energy Council (2006) *Global Wind Energy Outlook 2006*, http://www.gwec.net/uploads/media/GWEC_A4_0609_English.pdf; last accessed 18 March 2009.

Heier, S. (2006) *Grid Integration of Wind Energy Conversion Systems*, John Wiley & Sons, Ltd, Chichester, ISBN 10: 0470868996.

Holdsworth, L., Wu, X., Ekanayake, J. B. and Jenkins, N. (2003) Comparison of fixed speed and doubly-fed induction wind turbines during power system disturbances, *IEE Proceedings: Generation, Transmission and Distribution*, **150** (3), 343–352.

Manwell, J. F., McGowan, J. G. and Rogers, A. L. (2002) *Wind Energy Explained: Theory, Design and Application*, John Wiley & Sons, Ltd, Chichester, ISBN 10: 0471499722.

Müller, S., Deicke, M. and De Doncker, R. W. (2002) Doubly fed induction generator systems for wind turbines, *IEEE Industry Applications Magazine*, **8** (3), 26–33.

National Grid (2008) *GB Grid Code*, www.nationalgrid.com/uk/Electricity/Codes/gridcode/; last accessed 18 March 2009.

Slootweg, J. G. (2003) Wind power: modelling and impacts on power system dynamics. PhD thesis. Technical University of Delft.

UK HM Treasury (2006) *Stern Review on the Economics of Climate Change*, http://www.hm-treasury.gov.uk/sternreview_index.htm; last accessed 18 March 2009.

UCTE, Union for the Co-ordination of Transmission of Electricity (2004) Integrating wind power in the European power systems: prerequisites for successful and organic growth, UCTE Position Paper.

2

Power Electronics for Wind Turbines

Power electronic systems are frequently used for electrical power conversion at a wind turbine generator level, wind farm level or both. Within the wind turbine generator, power electronic converters are used to control the steady-state and dynamic active and reactive power flows to and from the electrical generator (Figure 2.1a).

Variable-speed wind turbines decouple the rotational speed from the grid frequency through a power electronic interface. Variable-speed operation can be achieved by using any suitable combination of generator, synchronous or asynchronous, and a power electronic interface. The commonly used variable-speed configurations are the fully rated converter (FRC) wind turbine, which may use synchronous or asynchronous generators, and the doubly fed induction generator (DFIG) wind turbine. In FRC wind turbines, the power electronic interface is connected to the stator of the generator. In DFIG wind turbines, the power electronic interface is connected between the stator and the rotor, and allows variable-speed operation of the wind turbine by injecting a variable voltage into the rotor at slip frequency.

In many cases, the electrical networks to which wind farms are connected are 'weak', with high source impedances. The output of a wind farm changes constantly with wind conditions and so causes variations in the voltage at the point of connection. This can be compensated by proper control of the power electronic converters of the wind turbine generators, within the rating and operating capability of the wind turbine generator and the converters. In addition, a power electronic reactive power compensator such as a static var compensator (SVC) or a static compensator (STATCOM) can be connected to the wind farm point of connection with the grid to supply reactive

Wind Energy Generation: Modelling and Control   Olimpo Anaya-Lara, Nick Jenkins,
Janaka Ekanayake, Phill Cartwright and Mike Hughes
© 2009 John Wiley & Sons, Ltd

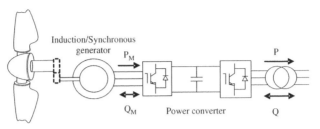

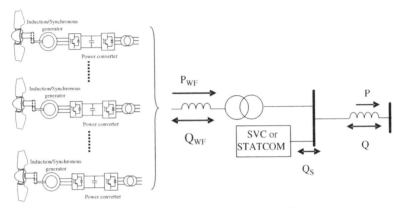

(a) Variable speed wind turbine

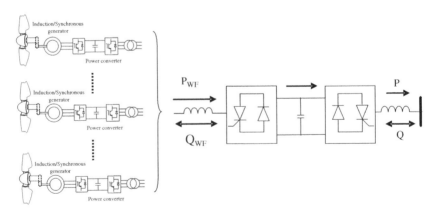

(b) Array of wind turbines connected to the AC network

(c) Array of wind turbines connected to the AC network through HVDC

Figure 2.1   Power electronic technologies for wind farms – active and reactive power flows.

power locally and thus improve the voltage profile at that point (Figure 2.1b). A reactive power compensator can also be used to improve the stability of the network to which the wind farm is connected. Effective voltage control through reactive power compensation facilitates the connection of medium- and large-sized wind turbines to 'weak' networks.

For large wind farms, high-voltage direct current (HVDC) power electronic converters may be employed to control bulk active power transfer through the DC link with independent control of reactive power both at the wind farm and to the network, as shown in Figure 2.1c.

2.1 Soft-starter for FSIG Wind Turbines

A soft-starter unit is used with FSIG wind turbines (Figure 1.6) to build up the magnetic flux slowly and so minimize transient currents during energization. The soft-starter consists of six thyristors, two per phase connected in anti-parallel, as shown in Figure 2.2a. An RC snubber circuit is connected in series across the thyristors to control the rate of change of voltage across the thyristors. A thyristor is latched in the on-state by a positive firing pulse supplied to the gate terminal with respect to the cathode, but must be turned off by natural commutation, that is, by reverse current flow in the power circuit. Therefore, an external controller is required to generate the on firing pulses to the gates (Th_F and Th_R). Automatic transition to the off-state occurs when the device current reaches zero.

To operate in current limiting mode, delayed firing pulses (Th_{Fa} to Th_{Rc}) are generated with increasing conduction period for each thyristor, as shown in Figure 2.2b. In this way, the phase current is gradually increased to the rated current. When the rated current is reached, the soft-starter is by-passed by the contactor.

The firing pulse generator for the soft-starter is shown in Figure 2.3. A phase-locked loop (PLL) is used to lock the firing pulses to the three-phase line voltages (A, B and C). This block returns an array of six saw-tooth phase signals separated $60°$ in phase to one another. The firing pulses are generated by comparing the saw-tooth waveforms with the defined firing characteristic (which is an increasing DC level).

2.2 Voltage Source Converters (VSCs)

2.2.1 The Two-level VSC

The two-level VSC is widely used in variable-speed drives and variable-speed wind turbine applications. The main advantages of the two-level VSC include

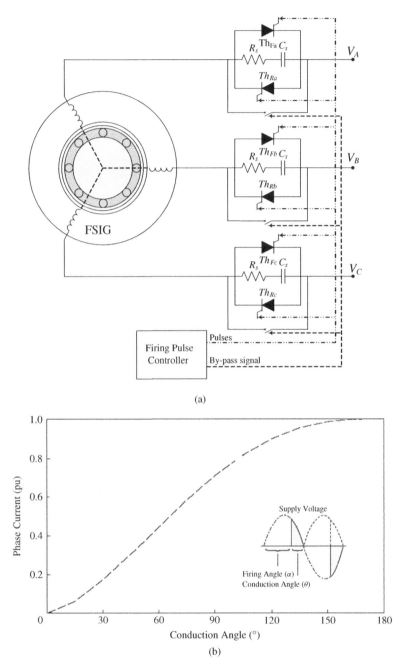

(a)

(b)

Figure 2.2   Soft-starter. (a) Configuration; (b) effect of firing and conduction angle in a thyristor

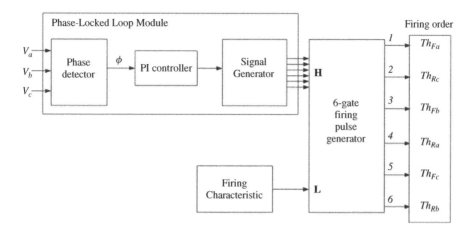

Figure 2.3   Control model of the soft-starter

its simplicity, proven technology and the possibility of building redundancy
into the string of series-connected switching devices, usually insulated gate
bipolar transistors (IGBTs). The two-level VSC allows the IGBTs to be connected in series, depending on the voltage rating of the device available and
the supply voltage required. The basic principle of a single-phase, two-level
VSC is shown in Figure 2.4, where it can be seen that the output waveform
has two levels, $+V_{DC}$ and $-V_{DC}$. Therefore, each switch string must be rated
for the full direct voltage, V_{DC}. Due to the large capacitance of the DC side
of the converter, the DC voltage, V_{DC}, is more or less constant and thus the
converter is known as a voltage source converter.

Three single-phase, two-level voltage source converters can be connected
to the same capacitor to form a three-phase converter. This converter power
circuit arrangement is often called the six-pulse converter configuration
(Figure 2.5). In this circuit, the switches in one leg are switched alternatively

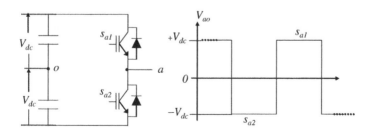

Figure 2.4   Fundamental principles of a single-phase, two-level converter

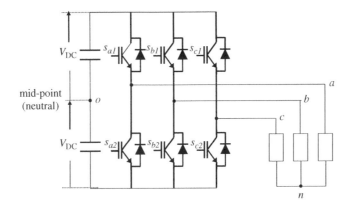

Figure 2.5   Three-phase, two-level voltage source converter

with a small dead time to avoid both conducting simultaneously. Therefore, one switching function is enough to control both switches in a leg.

There are a number of different switching strategies for VSCs (Mohan *et al.*, 1995; Boost and Ziogas, 1988; Holtz, 1992). These include square-wave operation, carrier-based pulse-width modulation (CB-PWM) techniques such as switching frequency optimal PWM (SFO-PWM), sinusoidal regular sampled PWM (RS-PWM), non-regular sampled PWM (NRS-PWM), selective harmonic elimination PWM (SHEM), space vector PWM (SV-PWM) and hysteresis switching techniques.

2.2.2   Square-wave Operation

In this technique (Figure 2.6), each switch conducts for almost 180°. No two switches in the same leg conduct simultaneously. Six patterns exist for one output cycle and the rate of sequencing these patterns specifies the bridge output

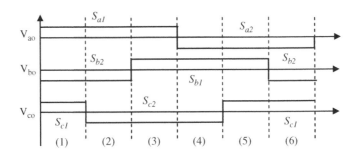

Figure 2.6   Three-phase output for fundamental frequency modulation

frequency. The six conducting switching patterns during six distinct intervals [marked as (1) to (6) in Figure 2.6] are $S_{c1}S_{b2}S_{a1}$, $S_{b2}S_{a1}S_{c2}$, $S_{a1}S_{c2}S_{b1}$, $S_{c2}S_{b1}S_{a2}$, $S_{b1}S_{a2}S_{c1}$ and $S_{a2}S_{c1}S_{b2}$.

With fundamental frequency switching, the switching losses are low (since switching losses are proportional to the switching frequency), but the harmonic content of the output waveforms is relatively high. The output voltage contains harmonics of the order $(6k \pm 1)$, where k is an integer.

2.2.3 Carrier-based PWM (CB-PWM)

This is the classical PWM where a reference signal, V_{ref}, which varies sinusoidally, is compared with a fixed-frequency triangular carrier waveform, V_{tri}, to create a switching pattern. If the single-phase two-level circuit of Figure 2.4 is considered with the waveforms shown in Figure 2.7, then S_{a1} is ON when $V_{ref} > V_{tri}$. S_{a2} is ON when $V_{ref} < V_{tri}$. In general:

$$S_{a1} = \begin{cases} 1 & V_{ref} > V_{tri} \\ 0 & V_{ref} < V_{tri} \end{cases} \tag{2.1}$$

$$S_{a2} = \begin{cases} 1 & V_{ref} < V_{tri} \\ 0 & V_{ref} > V_{tri} \end{cases} \tag{2.2}$$

where '1' denotes the switch state ON and '0' denotes the switch state OFF.

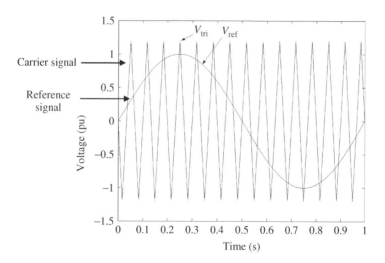

Figure 2.7   Reference voltage, V_{ref}, and the carrier waveform, V_{tri}

The amplitude modulation ratio, m_a, is defined as the ratio of the reference signal to the carrier signal:

$$m_a = \frac{\hat{V}_{ref}}{\hat{V}_{tri}} \qquad (2.3)$$

where the 'hat', '$\hat{\ }$', represents peak values.

The frequency modulation ratio, m_f, is defined as the ratio of the carrier frequency, f_{tri}, to the reference signal frequency, f_{ref}:

$$m_f = \frac{f_{tri}}{f_{ref}} \qquad (2.4)$$

It can be estimated from the waveforms in Figure 2.7 that $m_a = 0.8$ and $m_f = 15$.

The PWM switching pattern and the Fourier spectrum of this output waveform are shown in Figure 2.8.

If m_a is increased beyond unity ($m_a > 1.0$), then the fundamental voltage does not vary linearly. This condition is termed over-modulation (Mohan

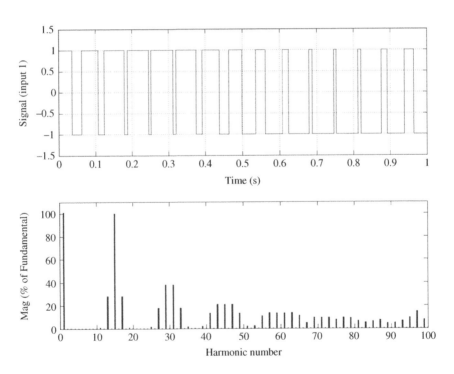

Figure 2.8   Output PWM waveform of a single-phase, two-level VSC and harmonic spectrum

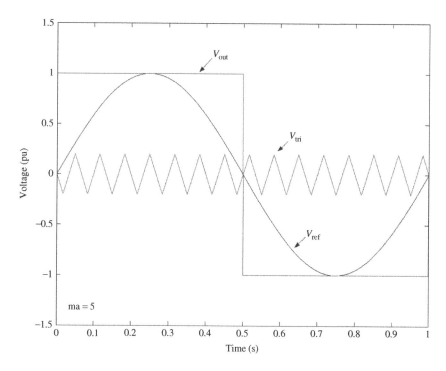

Figure 2.9 Over-modulation shown for a single-phase, two-level VSC

et al., 1995). As m_a is increased beyond 3.24, the output waveform degenerates into a square waveform (Figure 2.9).

This PWM switching strategy provides good results in terms of harmonic distortion, possibly eliminating the requirement for passive harmonic filtering but the switching losses are increased over square-wave modulation.

2.2.4 Switching Frequency Optimal PWM (SFO-PWM)

Higher utilization of the DC link can be achieved by using reference waveforms other than pure sinusoidal waveforms.

2.2.4.1 Trapezoidal Modulating Function

The reference signal is a trapezoidal function (Figure 2.10), which increases the ratio of the fundamental component of the maximum phase voltage to the DC supply voltage. This then reduces the ratings required for the converter elements and decreases the turn-on losses of the converter elements. However, the lower-order harmonic content of the output waveform is increased (Holtz, 1992; Taniguchi *et al.*, 1994).

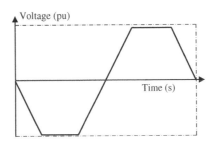

Figure 2.10   Trapezoidal modulating function

2.2.4.2   Third Harmonic Modulating Function

As shown in Figure 2.11 a third harmonic may be added to the reference sinusoidal waveform to increase the output fundamental frequency voltage and to allow an increase in m_a (Mohan *et al.*, 1995; Holtz, 1992).

2.2.5   *Regular and Non-regular Sampled PWM (RS-PWM and NRS-PWM)*

In CB-PWM methods, a reference signal, V_{ref}, whose waveform is required to be reproduced in the output waveform, is imposed on the carrier wave signal V_{tri}. As described by Bowes (1975), the equations describing the CB-PWM are not suitable for digital implementation. The need to implement PWM within digital or microprocessor-based systems led to the development of regular sampled (RS-PWM) and non-regular sampled (NRS-PWM) techniques.

In the RS-PWM, the reference signal is sampled at equidistant time instants (T_s). Figure 2.12 shows an example where two samples per carrier cycle are generated. The pulse width of the RS-PWM signal is modulated based on the magnitude of the samples h_1 and h_2. Several variations of the RS-PWM modulation technique can be found in the literature namely, 'single-edge',

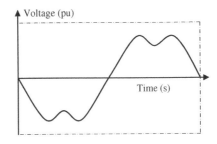

Figure 2.11   Harmonic modulating function

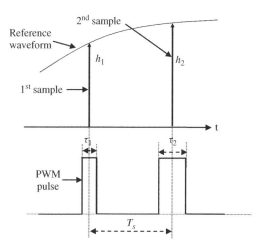

Figure 2.12   Regular sampling

'symmetrically double-edge' or 'asymmetrically double-edge' modulation (Bowes, 1975; Bowes and Lai, 1997).

In the 'non-regular sampling' or 'natural sampling' technique, the modulation time instants are not equidistant but are dependent on the modulation process.

2.2.6 Selective Harmonic Elimination PWM (SHEM)

This switching technique was described by Patel and Hoft (1973), where additional switching events are used. The values of the firing angle, α_k, are derived to eliminate particular harmonics. Figure 2.13 shows a periodic waveform from a two-level converter with an arbitrary number of chops per half cycle. Assuming that this periodic waveform has half-wave odd symmetry and the angle corresponding to the chop $\pi/2$ is α_k then from Fourier analysis:

$$V_{\mathrm{ao}} = \sum_{n=1}^{\infty} b_n \sin(n\omega t) \tag{2.5}$$

where

$$
\begin{aligned}
b_n &= \frac{2}{\pi} \int_0^{\pi} V_{\mathrm{ao}}(\omega t)\sin(n\omega t)\mathrm{d}(\omega t) \\[2mm]
&= \frac{V_{\mathrm{DC}}}{\pi}\left[
\begin{aligned}
&\int_0^{\alpha_1}\sin(n\omega t)\mathrm{d}(\omega t) - \int_{\alpha_1}^{\alpha_2}\sin(n\omega t)\mathrm{d}(\omega t) \\
&-\int_{\pi-\alpha_2}^{\pi-\alpha_1}\sin(n\omega t)\mathrm{d}(\omega t) + \int_{\pi-\alpha_1}^{\pi}\sin(n\omega t)\mathrm{d}(\omega t)
\end{aligned}
\right]
\end{aligned}
\tag{2.6}
$$

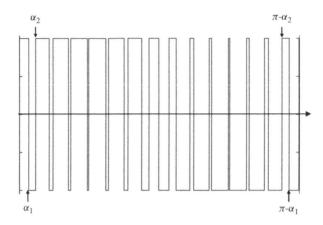

Figure 2.13   Two-level VSC with an arbitrary number of chops per half cycle

The integration given in Eq. (2.6) can be written as

$$b_n = \frac{4V_{DC}}{\pi n}\left[1 + 2\sum_{i=1}^{k}(-1)^i \cos(n\alpha_i)\right] \qquad (2.7)$$

for odd n.

By introducing k number of chops per half cycle into the converter output voltage (Figure 2.13), any k number of harmonics can be eliminated from the output voltage. The angles corresponding to the k chops can be found by equating $b_n = 0$ for the k harmonics to be eliminated and solving those k simultaneous equations for α_i (for $i = 1$ to k). As an example, if two chops were used with angles $\alpha_1 = 16.3°$ and $\alpha_2 = 22.1°$, the fifth and seventh harmonics can be eliminated from the output voltage waveform. However, there may be practical limitations in implementing this strategy (computational time and practically achievable firing angles). Hence a 'harmonic optimization' method has been proposed in (Buja and Indri, 1977) where a broader number of harmonics are adjusted to minimize the approximate RMS harmonic current of an induction motor.

2.2.7   Voltage Space Vector Switching (SV-PWM)

Another method of obtaining a pulse width modulation is based on space vector representation of the switching voltages in the $\alpha\beta$ plane as described by (Van der Broeck *et al*, 1988; Lindberg, 1990). This method has the advantage of being easier to implement than other PWM techniques and achieves similar results to regular sampled sinusoidal PWM with third harmonic added to the reference waveform. Therefore, the harmonic content of the output waveform is lower than that for an equivalent CB-PWM. A single rotating vector can be used to represent three-phase voltages. This vector is called the voltage space vector which generally rotates in a two-dimension plane. In this case, two stationary perpendicular axis, α and β, are used to represent the voltage space vector.

For a two-level converter using the space vector PWM switching strategy, the switching vectors are defined by the states of the converter switches as shown in Figure 2.14. Three switching legs, each having two states ON or OFF, allow the converter to produce ($2^3 = 8$) eight possible switching states (Figure 2.15).

In the SV-PWM technique, the sequence of the switching vectors is selected in such a way that only one leg is switched to move from one switching vector to the next. This switching sequence is achieved by arranging the adjacent active vectors and two-null vectors (Boost and Ziogas, 1988; Holtz, 1992). The switching times of the switching vectors are calculated by equating volt-second integrals between the required voltage vector and the switching vectors. Figure 2.16 shows an example for the calculation of switching times, when the required voltage vector is in the first sector.

The required voltage vector V_s is generated by using the adjacent active switching vectors V_1, V_2 and the null vectors V_0 and V_7. The times t_0 and

S_a	S_b	S_c	v_{ao}	v_{bo}	v_{co}	Switching vector
0	0	0	$-V_{DC}$	$-V_{DC}$	$-V_{DC}$	V_0
1	0	0	$+V_{DC}$	$-V_{DC}$	$-V_{DC}$	V_1
1	1	0	$+V_{DC}$	$+V_{DC}$	$-V_{DC}$	V_2
0	1	0	$-V_{DC}$	$+V_{DC}$	$-V_{DC}$	V_3
0	1	1	$-V_{DC}$	$+V_{DC}$	$+V_{DC}$	V_4
0	0	1	$-V_{DC}$	$-V_{DC}$	$+V_{DC}$	V_5
1	0	1	$+V_{DC}$	$-V_{DC}$	$+V_{DC}$	V_6
1	1	1	$+V_{DC}$	$+V_{DC}$	$+V_{DC}$	V_7

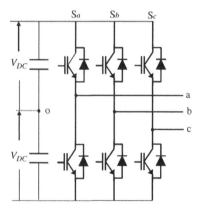

Figure 2.14 Switching status of a six-pulse converter switches

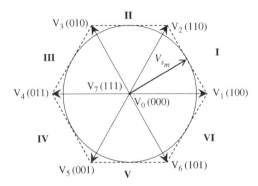

Figure 2.15 Switching vector positions for two-level SVPWM

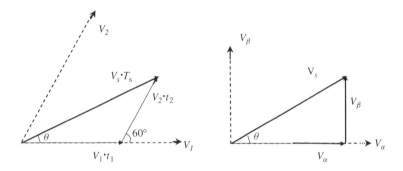

Figure 2.16 Phasor diagram and basic equations for switching time calculation

t_7 represent the switching duration of the null vectors V_0 and V_7 and t_1 and t_2 represent that for the active vectors V_1 and V_2. In a conventional space vector technique, the null vector switching times are chosen in such a way that $t_0 = t_7$. The magnitude of the required voltage vector is assumed to be constant during the switching period, T_s.

The switching duration of each vector is given by the following equations:

$$t_1 = T_s \frac{\sqrt{6}V_\alpha - V_\beta/\sqrt{2}}{2V_{DC}} \tag{2.8}$$

$$t_2 = \frac{T_s V_\beta}{\sqrt{2}V_{DC}} \tag{2.9}$$

and

$$t_0 = t_7 = \frac{T_s - t_1 - t_2}{2} \tag{2.10}$$

2.2.8 Hysteresis Switching

In this technique, the required converter output current is compared with the actual converter output current within a specified hysteresis band. If the actual current is more positive than the upper hysteresis level, then the converter is switched such that the current is reduced and vice versa (Brod and Novotny, 1985). Figure 2.17 shows the converter output current waveform for hysteresis switching.

One advantage of this control is its ease of implementation. However, this technique requires sufficient DC-link voltage to force the current in the desired direction. In addition, the converter switching frequency is influenced by several factors: ripple current magnitude, smoothing reactance, DC-link voltage and instantaneous grid voltage.

2.3 Application of VSCs for Variable-speed Systems

As discussed in Section 1.2, wind turbines use power electronic converters for variable-speed operation. Table 2.1 summarizes the application of VSCs for different generator configurations.

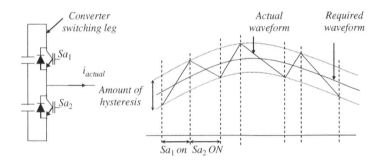

Figure 2.17 Hysteresis converter current control

Table 2.1 Generators and power electronics in wind turbine applications

Generator	Power electronic conversion used
DFIG	Back-to-back VSCs connected to the rotor
Permanent magnet synchronous generator-based FRC	Diode bridge-VSC or back-to-back VSCs connected to the armature
Wound rotor synchronous generator-based FRC	Diode bridge-VSC or back-to-back VSCs connected to the armature and field

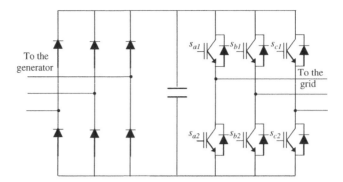

Figure 2.18 VSC with a three-phase diode bridge (this topology normally involves the use of a chopper)

2.3.1 VSC with a Diode Bridge

Figure 2.18 shows an arrangement of a VSC with a three-phase diode bridge. In this configuration, power can only flow from the generator to the grid. The generated AC voltage and current (and thus power) is converted into DC using the diode bridge and then inverted back to 50 Hz AC using the VSC. This arrangement decouples the wind turbine from the AC grid, thus allowing variable-speed operation of the wind turbine. The VSC is generally controlled to maintain the DC link capacitor at a constant voltage. This will ensure the power transfer between the DC link and the grid (if incoming power is not transferred to the grid, then the DC link voltage will increase). In wind turbine applications, the DC link may include a series inductor to form a filter with the capacitor to minimize the DC ripple, and a DC chopper to protect the DC link from over-voltages (in the case of an AC side fault).

2.3.2 Back-to-Back VSCs

The back-to-back VSC is a bi-directional power converter consisting of two voltage source converters as shown in Figure 2.19.

The IGBTs on the generator-side VSC are controlled using a PWM technique (usually based on SVPWM). In variable-speed wind turbines, the frequency of the reference sinusoidal waveform (V_{ref} in Figure 2.7) is locked to the frequency of the generated voltage. Therefore, the frequency of the output voltage of the VSC contains a component at the frequency of the generated voltage, referred to as the fundamental and also higher-order harmonics. The magnitude of the VSC output voltage can be controlled by changing the amplitude modulation index and the phase angle can be controlled by controlling the phase angle of V_{ref} with respect to the generated voltage.

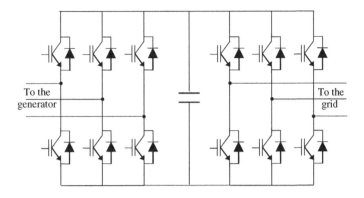

Figure 2.19   Back-to-back VSCs

In order to describe the operation of the VSC connected to the generator, it is assumed that the VSC produces a sinusoidal waveform (higher-order harmonic components are neglected). As the wind turbine generator can be represented by a voltage behind a reactance (Kundur, 1994), the generator-side connection of the VSC can be represented by the equivalent circuit shown in Figure 2.20, where V_G is the magnitude of the generated voltage, V_{VSC} is the magnitude of the VSC output voltage, δ is the phase angle between these two voltages and X_G is the equivalent generator reactance.

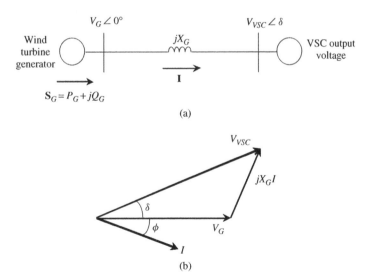

Figure 2.20   Active and reactive power transfer between the generator and the VSC. (a) Equivalent circuit diagram; (b) phasor diagram (Kundur, 1994)

The active power, P_G, and reactive power, Q_G, transferred from the generator to the VSC are defined as follows (Kundur, 1994):

$$P_G = \frac{V_G V_{VSC}}{X_G} \sin \delta \tag{2.11}$$

$$Q_G = \frac{V_G^2}{X_G} - \frac{V_G V_{VSC}}{X_G} \cos \delta \tag{2.12}$$

If the load angle δ is assumed to be small, then $\sin \delta \approx \delta$ and $\cos \delta \approx 1$. Hence Eqs (2.11) and (2.12) can be simplified to

$$P_G = \frac{V_G V_{VSC}}{X_G} \delta \tag{2.13}$$

$$Q_G = V_G \left(\frac{V_G - V_{VSC}}{X_G} \right) \tag{2.14}$$

From Eqs (2.13) and (2.14), it is seen that the active power transfer depends mainly on the load angle δ and the reactive power transfer depends mainly on the difference in voltage magnitudes. As V_{VSC} and δ can be controlled independently of the generator voltage, the VSC control facilitates control of the magnitude and the direction of the active and reactive power flow between the generator and the DC link. Similarly, as the other VSC is connected to the grid via a reactor or via a transformer, the power transfer between that VSC and the grid can also be described using the same principle.

Even though the active and reactive power transfers can be controlled by controlling V_{VSC} and δ, in practice to control the power transfer other parameters may be used. For example, by maintaining the DC link voltage constant, it is possible to make sure that the generated power is transferred to the grid.

The main advantages of using back-to-back VSCs include the following: (a) it is a well-established technology and has been used in machine drive-based applications for many years; (b) many manufacturers produce components especially designed for this type of converter; and (c) the decoupling of the two VSCs through a capacitor allows separate control of the two converters.

References

Boost, M. A. and Ziogas, P. D. (1988) State of the art carrier PWM techniques: a critical evaluation. *IEEE Transactions on Industry Applications*, **24** (2), 271–280.

Bowes, S. R. (1975) New sinusoidal PWM inverter, *Proceedings of the IEE*, **122** (11), 1279–1285.

Bowes, S. R. and Lai, Y. S. (1997) The relationship between space-vector modulation and regular sampled PWM, *IEEE Transactions on Industrial Electronics*, **44** (5), 670–679.

Brod, D. M. and Novotny, D. W. (1985) Current control of VSI-PWM inverters, *IEEE Transactions on Industrial Applications*, **1A-21** (4), 562–570.

Buja, G. S. and Indri, G. B. (1977) Optimal pulsewidth modulation for feeding AC motors, *IEEE Transactions on Industry Applications*, **1A-13** (1), 38–44.

Holtz, J. (1992) Pulsewidth modulation – a survey, *IEEE Transactions on Industrial Electronics*, **39** (5), 410–420.

Kundur, P. (1994) *Power System Stability and Control*, McGraw-Hill, New York, ISBN 0-07-035958-X.

Lindberg, L. (1990) voltage sourced converters for high power transmission applications. PhD Thesis. Royal Institute of Technology, Stockholm.

Mohan, N., Undeland, T. M. and Robbins, W. P. (1995) *Power Electronics – Converters, Applications and Design*, 2nd edn, John Wiley & Sons, Inc., New York, ISBN 0-471-58408-8.

Patel, H. S. and Hoft, R. G. (1973) Generalised techniques of harmonic elimination and voltage control in thyristor inverters: Part 1 – harmonic elimination, *IEEE Transactions on Industry Applications*, **1A-9** (1), 310–317.

Taniguchi, K., Inoue, M., Takeda, Y. and Morimoto, S. (1994) A PWM strategy for reducing torque-ripple in inverter-fed induction motor, *IEEE Transactions on Industry Applications*, **30** (1), 71–77.

Van der Broeck, H. W., Skudelny, H. C. and Stanke, G. V. (1988) Analysis and realization of a pulsewidth modulator based on voltage space vectors, *IEEE Transactions on Industry Applications*, **24** (1), 142–150.

3

Modelling of Synchronous Generators

3.1 Synchronous Generator Construction

A synchronous generator consists of two elements: the field and the armature. The field is located on the rotor and the armature on the stator. The armature has a concentrated three-phase winding as shown in Figure 3.1. The field winding carries direct current and produces a magnetic field which rotates with the rotor. The rotor of a low-speed generator, such as a hydro-turbine, has a non-uniform air-gap with a concentrated field winding as shown in Figure 3.1 and is referred to as a salient-pole generator. The rotor of a high-speed generator, used with steam and gas turbines, has a uniform air-gap with a distributed field winding and is referred to as a round-rotor (cylindrical pole) generator.

Even though in this chapter the dynamic equations are derived for a salient-pole generator, they are equally true for a round-rotor generator. Both produce a sinusoidal magnetic field in the air-gap. In the case of a salient-pole generator, the shape of the poles is formed to obtain a sinusoidal air-gap flux. In the case of round-rotor generators, the rotor windings are distributed over two-thirds of the rotor surface and the flux produced by them aggregates into a sinusoidal shape. Further, stator windings are also arranged to help produce a sinusoidal voltage waveform.

Hence, for modelling purposes, the only difference is in the parameter values due to the different physical constructions.

3.2 The Air-gap Magnetic Field of the Synchronous Generator

The concentrated stator windings of three phases a, b and c are represented by three equivalent windings $a - a', b - b'$ and $c - c'$ (Figure 3.2). When

Wind Energy Generation: Modelling and Control   Olimpo Anaya-Lara, Nick Jenkins,
Janaka Ekanayake, Phill Cartwright and Mike Hughes
© 2009 John Wiley & Sons, Ltd

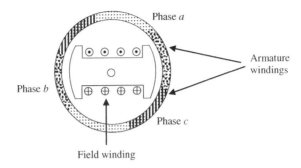

Figure 3.1   Schematic diagram of a three-phase synchronous generator

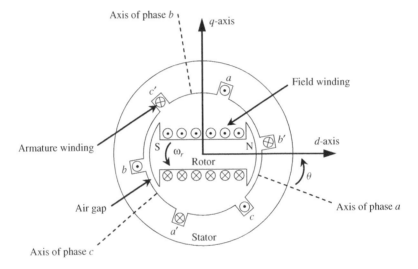

Figure 3.2   Schematic diagram of a three-phase synchronous generator (Kundur, 1994)

the rotor is driven by a prime mover, the magnetic field produced by the field winding rotates in space at synchronous speed ω_s. This magnetic field cuts the stator conductors and three voltages, which are displaced by 120° (in time), are induced in the three windings $a - a', b - b'$ and $c - c'$. If these windings are connected to three identical loads, the resulting three phase currents are also displaced by 120°, as shown in Figure 3.3. These currents will, in turn, each produce a magnetic field and the resultant magnetic field in the air-gap is the combination of the magnetic fields produced by the stator currents (referred to as the stator magnetic field, ϕ_s) and the magnetic field produced by the field winding (referred to as the rotor magnetic field, ϕ_r). For simplicity of analysis,

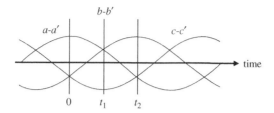

Figure 3.3   Three phase currents in the stator windings $a-a'$, $b-b'$ and $c-c'$

these two magnetic fields are considered separately and superposition is used
to obtain the resultant air-gap magnetic field.

As shown in Figure 3.3, when $t = 0$ the current in phase a is at its positive
maximum (I_m) and the currents in phases b and c are at their negative half
maxima ($-I_m/2$). If the effective number of turns of each phase of the stator
winding is N, then the current in phase a produces a component of the stator
magnetic field, ϕ_a, where the magnitude is proportional to the number of
ampere-turns, $N I_m$,[1] along the axis of phase a (see Figure 3.2). Similarly, the
currents in phases b and c produce two components of the stator magnetic field,
ϕ_b and ϕ_c, whose magnitudes are proportional to the number of ampere-turns,
$N I_m/2$ along the axes of phases b and c, respectively. These three magnetic
fields and the resultant stator magnetic field at $t = 0$ are shown in Figure 3.4a.

At time $t = t_1$ the currents in phases a and b produce two magnetic fields
whose magnitudes are proportional to the number of ampere-turns, $N I_m/2$
along the axes of phases a and b, respectively, and the current in phase c
produces a magnetic field whose magnitude is proportional to the number of
ampere-turns, $N I_m$, along the axis of phase c. The resultant stator magnetic
field at $t = t_1$ is then shifted by $\pi/3$ and is shown in Figure 3.4b. Similarly
at time $t = t_2$, the stator magnetic field further shifts by $\pi/3$ as shown in
Figure 3.4c.

---

[1] Flux density, B, is related to field intensity, H, by (Hindmarsh and Renfrew, 1996)

$$B = \mu \times H = \mu \times \frac{I \times N}{l}$$

where l is the length of the magnetic circuit.
On multiplying both sides by the cross-sectional area, A:

$$\phi = B \times A = \mu \times \frac{I \times N}{l} \times A = I \times N \times \frac{\mu \times A}{l} = I \times N \times \Lambda$$

where Λ is the reciprocal of the reluctance of the magnetic circuit.

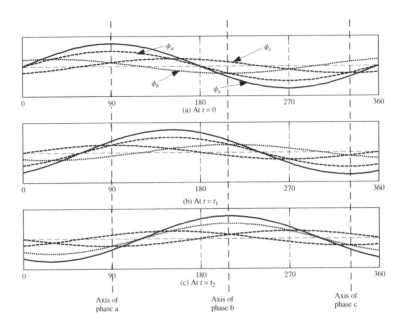

Figure 3.4   Stator magnetic field due to currents in three-phase windings

From Figure 3.4, it is clear that in each of the two time intervals $t_1 - 0 = \pi/3\omega_s$ and $t_2 - t_1 = \pi/3\omega_s$, the stator magnetic field has rotated by $\pi/3$. In other words, the field has rotated at the synchronous speed, ω_s. The peak value of the stator magnetic field is proportional to $3NI_m/2$.

A component of the stator magnetic field, ϕ_s, will link with the component of the rotor magnetic field, ϕ_r at the air-gap. The resultant magnetic field in the air-gap is then given by the vector sum of these magnetic fields. The components which are not contributing to the air-gap magnetic field are called leakage fluxes.

3.3   Coil Representation of the Synchronous Generator

It may be seen that with balanced three-phase currents, the three stator windings can be replaced by a single coil aligned with the axis of phase a, which carries a current of $3I_m/2$ and rotates at synchronous speed. This representation is often used for steady-state studies of electrical machines. Under dynamic conditions, as the currents in the three-phase windings change in magnitude and phase, the position of the resultant stator field vector changes and the

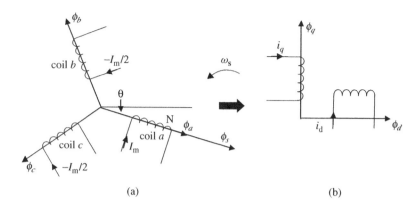

(a)   (b)

Figure 3.5 Three-phase to two-phase transformation: (a) three-coil representation; (b) two-coil representation

single-coil representation is no longer suitable. Therefore, for dynamic studies the electrical machine model is based on a two-phase representation.

Consider three coils, a, b and c, each carrying direct currents $I_m \cos 0 = I_m$, $I_m \cos(0 - 2\pi/3) = -I_m/2$ and $I_m \cos(0 - 4\pi/3) = -I_m/2$,[2] respectively, They rotate at a speed ω_s as shown in Figure 3.5a. The resultant magnetic field produced by three-phase windings will be proportional to $3NI_m/2$ in the direction of the axis of coil a. As time elapses, the magnitude of this magnetic field remains the same but rotates at synchronous speed, ω_s (Figure 3.5). Therefore, this three-coil structure fed with direct current and rotating at synchronous speed can be used as an analogue for the stator of a synchronous generator.

To define the two-phase system, two orthogonal coils are selected, one placed on the d axis, which is chosen to align with the rotor field winding position, and the other on the q axis, that leads the d axis by 90° (Figure 3.5b).

Resolving the resultant magnetic field produced by the three-phase windings, ϕ_s aligned with phase a, into the direction of d and q, Eqs (3.1) and (3.2) are obtained:

$$\phi_d = \phi_s \cos \theta \tag{3.1}$$

$$\phi_q = -\phi_s \sin \theta \tag{3.2}$$

---

[2] In the figure, the three coils are placed such that each coil makes 0 rad with the corresponding axes at $t = 0$, that is, coil a is placed on the axis of phase a, coil b on the axis of phase b and coil c on the axis of phase c.

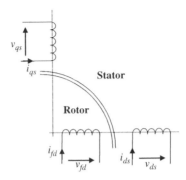

Figure 3.6 Two-coil representation of the synchronous generator

If the number of turns in the two-phase windings is N', the corresponding current relationships can be derived from Eqs (3.1) and (3.2) as

$$N' i_d = \frac{3}{2} N I_m \cos \theta \qquad\qquad (3.3)$$

$$N' i_q = -\frac{3}{2} N I_m \sin \theta \qquad\qquad (3.4)$$

Various authors select the ratio N/N' either as $2/3$ or as $\sqrt{2/3}$ (Fitzgerald *et al.*, 1992; Kundur, 1994; Krause *et al.*, 2002). In this book, N/N' is selected as $\sqrt{2/3}$.[3]

For a viewer on a platform which is rotating at synchronous speed (the synchronous rotating reference frame), the fluxes in the synchronous generator can be described by three stationary coils, two representing the stator field and one representing the rotor field (Figure 3.6). The stator coils d and q carry direct currents of $\sqrt{3/2} I_m \cos \theta$ and $-\sqrt{3/2} I_m \sin \theta$ respectively and the rotor coil carries the DC field current.

3.4   Generator Equations in the dq Frame

Before deriving the generator equations for the synchronous machine, consider two mutually coupled stationary coils:

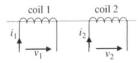

<hr />

[3] When N/N' is selected as $\sqrt{2/3}$ the power calculated in the dq coordinate system is the same as that in the *abc* system and therefore called the *power-invariant dq* transformation.
If $N/N' = 3/2$ is used instead, then the dq transformation is said to be *amplitude-invariant*.

The flux associated with coil 1 may be expressed as (Krause *et al.*, 2002)

$$\phi_1 = \phi_{l1} + \phi_{m1} + \phi_{m2} \tag{3.5}$$

where ϕ_{l1} is the leakage flux due to coil 1, ϕ_{m1} is the flux between coils 1 and 2 due to the current in coil 1 and ϕ_{m2} is the flux between coils 1 and 2 due to the current in coil 2.

The voltage equation for coil 1 can be expressed as

$$v_1 = r_1 i_1 + N_1 \frac{d\phi_1}{dt} = r_1 i_1 + \frac{d\psi_1}{dt} \tag{3.6}$$

where r_1 is the resistance of the coil 1, N_1 is the number of turns in coil 1 and ψ_1 is the flux linkage[4] with coil 1.

For modelling purposes, it is convenient to express flux linkage in terms of inductance and currents. From Eqs (3.5) and (3.6), the flux linking with coil 1 can be written as (Krause *et al.*, 2002)

$$\psi_1 = L_{l1} \times i_1 + L_m \times i_1 + L_m \times i_2 \tag{3.7}$$

where L_{l1} is the leakage inductance of coil 1 and L_m is the mutual inductance between coils 1 and 2. The term $L_{l1} + L_m$ which is associated with coil 1 is generally referred to as the self-inductance and L_m is referred to as the mutual inductance.

The self- and mutual inductances, which govern the voltage equations of the synchronous generator, vary with angle θ, which in turn varies with time. However, in the synchronously rotating reference frame, both stator and rotor fluxes are seen as stationary. Hence the flux linkage and thus inductances are constant. The generator stator and rotor equations in the dq reference frame, where the d axis is oriented with the field flux vector and the q axis is assumed to be 90° ahead of the d axis in the direction of rotation, are given in Table 3.1 (Kundur, 1994; Krause *et al.*, 2002). When deriving these equations it was assumed that the three-phase currents are balanced.

The voltage equations Eqs (3.8) and (3.9) are very similar to the voltage equation Eq. (3.6) derived for two stationary coils. However an additional term of speed is present in these equations. This term results from the transformation into the synchronous reference frame and is referred to as 'speed voltage' (due to flux changes in space) (Kundur, 1994; Krause *et al.*, 2002). The 'speed voltage' term does not appear in the rotor voltage equation as the field coil is stationary in the synchronous reference frame.

---

[4] Flux linkage = number of turns × flux ($\psi = N \times \phi$).

Table 3.1 Synchronous generator equations in the dq domain

Stator voltage equations:		Stator flux equations:	
$v_{ds} = -r_s i_{ds} - \omega_s \psi_{qs} + \dfrac{d}{dt}\psi_{ds}$	(3.8)	$\psi_{ds} = -L_{ls}i_{ds} + L_{md}(-i_{ds} + i_f)$	(3.10)
$v_{qs} = -r_s i_{qs} + \omega_s \psi_{ds} + \dfrac{d}{dt}\psi_{qs}$	(3.9)	$\psi_{qs} = -L_{ls}i_{qs} + L_{mq}(-i_{qs})$	(3.11)
Rotor voltage equations:		**Rotor flux equations:**	
$v_f = r_f i_f + \dfrac{d}{dt}\psi_f$	(3.12)	$\psi_f = L_{lf}i_f + L_{md}(-i_{ds} + i_f)$	(3.13)

In order to obtain a per unit (pu) representation of the voltage equations, consider Eq. (3.8). Dividing Eq. (3.8) by the base value of impedance, Z_b, given by $Z_b = \frac{V_b}{I_b} = \omega_b L_b$:

$$\frac{v_{ds}}{Z_b} = -\frac{r_s}{Z_b}i_{ds} - \frac{\omega_s \psi_{qs}}{Z_b} + \frac{1}{Z_b}\frac{d}{dt}\psi_{ds}$$
$$\frac{v_{ds}}{V_b} = -\frac{r_s}{Z_b}\frac{i_{ds}}{I_b} - \frac{\omega_s}{\omega_b}\frac{\psi_{qs}}{I_b L_b} + \frac{1}{\omega_b}\frac{d}{dt}\frac{\psi_{ds}}{I_b L_b} \tag{3.14}$$

As the base value of the flux linkage is given by $\psi_b = L_b I_b$, Eq. (3.14) can be represented by

$$\bar{v}_{ds} = -\bar{r}_s \bar{i}_{ds} - \bar{\omega}_s \bar{\psi}_{qs} + \frac{1}{\omega_b}\frac{d}{dt}\bar{\psi}_{ds} \tag{3.15}$$

with pu quantities represented by an upper bar.

Table 3.2 Synchronous generator equations in the dq domain (in pu)

Stator voltage equations:		Stator flux equations:	
$\bar{v}_{ds} = -\bar{r}_s \bar{i}_{ds} - \bar{\omega}_s \bar{\psi}_{qs} + \dfrac{1}{\omega_b}\dfrac{d}{dt}\bar{\psi}_{ds}$	(3.16)	$\bar{\psi}_{ds} = -\bar{L}_{ls}\bar{i}_{ds} + \bar{L}_{md}(-\bar{i}_{ds} + \bar{i}_f)$	(3.18)
$\bar{v}_{qs} = -\bar{r}_s \bar{i}_{qs} + \bar{\omega}_s \bar{\psi}_{ds} + \dfrac{1}{\omega_b}\dfrac{d}{dt}\bar{\psi}_{qs}$	(3.17)	$\bar{\psi}_{qs} = -\bar{L}_{ls}\bar{i}_{qs} + \bar{L}_{mq}(-\bar{i}_{qs})$	(3.19)
Rotor voltage equations:		**Rotor flux equations:**	
$\bar{v}_f = \bar{r}_f \bar{i}_f + \dfrac{1}{\omega_b}\dfrac{d}{dt}\bar{\psi}_f$	(3.20)	$\bar{\psi}_f = \bar{L}_{lf}\bar{i}_f + \bar{L}_{md}(-\bar{i}_{ds} + \bar{i}_f)$	(3.21)

In Eq. (3.15), all quantities are in pu except time, which is in seconds and base angular frequency, which is in radians per second. Similarly, all the other equations in Table 3.1 are represented by pu quantities as shown in Table 3.2.

When the synchronous generator carries unbalanced currents, the zero sequence current component, $\bar{i}_{0s}$, should also be considered. Under such conditions, in addition to the equations given in Table 3.2, the following equations should also be considered:

$$\bar{v}_{0s} = -\bar{r}_s\bar{i}_{0s} + \frac{1}{\omega_b}\frac{d}{dt}\bar{\psi}_{0s} \tag{3.22}$$

$$\bar{\psi}_{0s} = -\bar{L}_{ls}\bar{i}_{0s} \tag{3.23}$$

3.4.1 Generator Electromagnetic Torque

The generator torque is given by the cross-product of the stator flux and stator current:

$$T_e = \bar{\psi}_{ds} \cdot \bar{i}_{qs} - \bar{\psi}_{qs} \cdot \bar{i}_{ds} \tag{3.24}$$

3.5 Steady-state Operation

Under steady-state conditions, the d/dt terms in Eqs (3.16), (3.17) and (3.20) are equal to zero. With $\bar{L}_d = \bar{L}_{ls} + \bar{L}_{md}$, $\bar{L}_q = \bar{L}_{ls} + \bar{L}_{mq}$ and $\bar{L}_f = \bar{L}_{lf} + \bar{L}_{md}$, Eqs (3.16)–(3.21) can be reduced as given in Table 3.3.

Table 3.3 Synchronous generator reduced equations in the dq domain (in pu)

Stator voltage equations:		Stator flux equations:	
$\bar{v}_{ds} = -\bar{r}_s\bar{i}_{ds} - \bar{\omega}_s\bar{\psi}_{qs}$	(3.25)	$\bar{\psi}_{ds} = -\bar{L}_{ds}\bar{i}_{ds} + \bar{L}_{md}\bar{i}_f$	(3.27)
$\bar{v}_{qs} = -\bar{r}_s\bar{i}_{qs} + \bar{\omega}_s\bar{\psi}_{ds}$	(3.26)	$\bar{\psi}_{qs} = -\bar{L}_{qs}\bar{i}_{qs}$	(3.28)
Rotor voltage equations:		**Rotor flux equations:**	
$\bar{v}_f = \bar{r}_f\bar{i}_f$	(3.29)	$\bar{\psi}_f = \bar{L}_l\bar{i}_f - \bar{L}_{md}\bar{i}_{ds}$	(3.30)

Substituting for flux terms in Eqs (3.25) and (3.26) from Eqs (3.27) and (3.28), the following two equations can be obtained:

$$\bar{v}_{ds} = -\bar{r}_s\bar{i}_{ds} + \bar{\omega}_s\bar{L}_{qs}\bar{i}_{qs} = -\bar{r}_s\bar{i}_{ds} + \bar{X}_{qs}\bar{i}_{qs} \tag{3.31}$$

$$\overline{v}_{qs} = -\overline{r}_s\overline{i}_{qs} - \overline{\omega}_s\overline{L}_{ds}\overline{i}_{ds} + \overline{\omega}_s\overline{L}_{md}\overline{i}_f = -\overline{r}_s\overline{i}_{qs} - \overline{X}_{ds}\overline{i}_{ds} + \overline{\omega}_s\overline{L}_{md}\overline{i}_f \quad (3.32)$$

where $\overline{X}_{qs} = \overline{\omega}_s\overline{L}_{qs}$ and $\overline{X}_{ds} = \overline{\omega}_s\overline{L}_{ds}$.

From Eq. (3.29), $\overline{i}_f$ in Eq. (3.32) can be replaced by $\overline{v}_f/\overline{r}_f$ and with the definition of $\overline{E}_{fd} = \overline{\omega}_s\overline{L}_{md}\overline{v}_f/\overline{r}_f$, we obtain

$$\overline{v}_{qs} = -\overline{r}_s\overline{i}_{qs} - \overline{X}_{ds}\overline{i}_{ds} + \overline{E}_{fd} \quad (3.33)$$

The armature terminal voltage is expressed as $\overline{E}_t = \overline{v}_{ds} + j\overline{v}_{qs}$ and from Eqs (3.31) and (3.33), the following steady-state equation of the synchronous machine can be obtained:

$$\overline{E}_t = \overline{v}_{ds} + j\overline{v}_{qs} = -\overline{r}_s(\overline{i}_{ds} + j\overline{i}_{qs}) + (\overline{X}_{qs}\overline{i}_{qs} - j\overline{X}_{ds}\overline{i}_{ds}) + j\overline{E}_{fd} \quad (3.34)$$

If saliency is neglected, $\overline{X}_{qs} = \overline{X}_{ds} = \overline{X}_s$ and, with $\overline{I}_t = \overline{i}_{ds} + j\overline{i}_{qs}$ Eq. (3.34) can be reduced to

$$\begin{aligned}
\overline{E}_t &= -\overline{r}_s\overline{I}_t + \overline{X}_s(\overline{i}_{qs} - j\overline{i}_{ds}) + j\overline{E}_{fd} \\
&= -\overline{r}_s\overline{I}_t + \overline{X}_s(-j^2\overline{i}_{qs} - j\overline{i}_{ds}) + j\overline{E}_{fd} \quad (3.35) \\
&= -(\overline{r}_s + j\overline{X}_s)\overline{I}_t + j\overline{E}_{fd}
\end{aligned}$$

Equation (3.35) defines the steady-state equation of the synchronous machine and can be represented by the phasor diagram shown in Figure 3.7a and the equivalent circuit shown in Figure 3.7b.

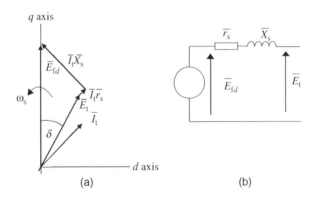

(a)   (b)

Figure 3.7   Phasor diagram and the equivalent circuit for steady-state operation

3.6   Synchronous Generator with Damper Windings

In both salient-pole and cylindrical-pole generators, solid copper bars run through the rotor to provide additional paths for circulating damping currents. In the case of salient-pole generators, damper bars are set into the pole faces as shown in Figure 3.8a. Even though discontinuous end-rings are shown in the figure, in some machines a continuous end-ring exists which links poles thus providing a further and major path for q axis damper current flow. In the case of cylindrical-pole generators, the coil wedges shown in Figure 3.8b are connected together by end-rings to form damping circuits. However, for clarity the end rings are not shown in the figure.

The currents in the damper windings interact with the air-gap flux and produce a torque which provides damping of rotor oscillations following a transient disturbance. In the case of an unsymmetrical fault on the network to which the synchronous machine is connected, the air-gap flux would have two components, positive sequence (flux due to current in the direction of rotation) and negative sequence (flux due to currents in the reverse direction of rotation). The opposing directions of rotor and negative sequence flux give a high relative speed and hence a large torque contribution.

The currents in the damper windings interact with the negative sequence air-gap flux and produce a counteracting torque which reduces the accelerating torque, thus limiting the rate of increase of the machine speed.

The currents in the damper windings can be resolved into two components. The circulating damping current under a pole forms the d axis damping current;

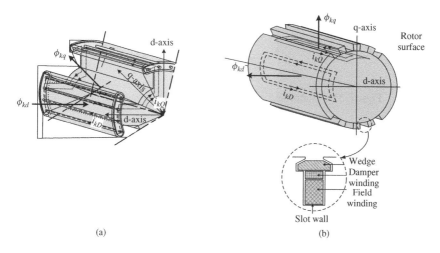

(a)   (b)

Figure 3.8   Damper windings and circulating current paths: (a) for a salient-pole generator; (b) for a cylindrical-pole generator

whereas the circulating damping current between two pole faces forms the q axis damping current (Figure 3.8). In the generator model shown in Figure 3.9, these currents were assumed to flow in sets of closed circuits: one set whose flux is in line with that of the field along the d axis and the other set whose flux is along the q axis. In the simplified model representing the synchronous generator, only one damper winding along the q axis is used, but often two damper windings, kq_1 and kq_2, are represented (Figure 3.9). Although the same basic representation can be used for both salient-pole and cylindrical-pole generators, the circuit parameters representing the damper windings are widely different.

The synchronous generator equations in the dq domain are summarized in Table 3.4.

The stator voltage equations Eqs (3.36) and (3.37) and the rotor voltage equations Eqs (3.40)–(3.43) are written in terms of currents and flux linkages. The flux linkages and the currents are related and both cannot be independent.

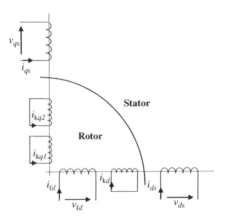

Figure 3.9 Stator and rotor circuits of a synchronous generator

Table 3.4 Synchronous generator equations in the dq domain including damper windings

Stator voltage equations:	Stator flux equations:
$$\bar{v}_{ds} = -\bar{r}_s \bar{i}_{ds} - \bar{\omega}_s \bar{\psi}_{qs} + \frac{1}{\omega_b}\frac{d}{dt}\bar{\psi}_{ds} \quad (3.36)$$	$$\bar{\psi}_{ds} = -\bar{L}_{ls}\bar{i}_{ds} + \bar{\psi}_{md} \quad (3.38)$$
$$\bar{v}_{qs} = -\bar{r}_s \bar{i}_{qs} + \bar{\omega}_s \bar{\psi}_{ds} + \frac{1}{\omega_b}\frac{d}{dt}\bar{\psi}_{qs} \quad (3.37)$$	$$\bar{\psi}_{qs} = -\bar{L}_{ls}\bar{i}_{qs} + \bar{\psi}_{mq} \quad (3.39)$$

(continued overleaf)

Table 3.4 *(continued)*

Rotor voltage equations:		Rotor flux equations:	
$\bar{v}_{fd} = \bar{r}_{fd}\bar{i}_{fd} + \dfrac{1}{\omega_b}\dfrac{d}{dt}\bar{\psi}_{fd}$	(3.40)	$\bar{\psi}_{fd} = \bar{L}\,\bar{i}_{fd} + \bar{\psi}_{md}$	(3.44)
$\bar{v}_{kd} = \bar{r}_{kd}\bar{i}_{kd} + \dfrac{1}{\omega_b}\dfrac{d}{dt}\bar{\psi}_{kd}$	(3.41)	$\bar{\psi}_{kd} = \bar{L}_{lkd}\bar{i}_{kd} + \bar{\psi}_{md}$	(3.45)
$\bar{v}_{kq1} = \bar{r}_{kq1}\bar{i}_{kq1} + \dfrac{1}{\omega_b}\dfrac{d}{dt}\bar{\psi}_{kq1}$	(3.42)	$\bar{\psi}_{kq1} = \bar{L}_{lkq1}\bar{i}_{kq1} + \bar{\psi}_{mq}$	(3.46)
$\bar{v}_{kq2} = \bar{r}_{kq2}\bar{i}_{kq2} + \dfrac{1}{\omega_b}\dfrac{d}{dt}\bar{\psi}_{kq2}$	(3.43)	$\bar{\psi}_{kq2} = \bar{L}_{lkq2}\bar{i}_{kq2} + \bar{\psi}_{mq}$	(3.47)

where

$$\bar{\psi}_{md} = \bar{L}_{md}(-\bar{i}_{ds} + \bar{i}_{fd} + \bar{i}_{kd})$$

$$\bar{\psi}_{mq} = \bar{L}_{mq}(-\bar{i}_{qs} + \bar{i}_{kq1} + \bar{i}_{kq2})$$

3.7 Non-reduced Order Model

The currents in terms of flux linkages are obtained from Eqs (3.38) and (3.39) and Eqs (3.44)–(3.47) and given as follows:

$$\bar{i}_{ds} = -\frac{1}{\bar{L}_{ls}}(\bar{\psi}_{ds} - \bar{\psi}_{md}) \tag{3.48}$$

$$\bar{i}_{qs} = -\frac{1}{\bar{L}_{ls}}(\bar{\psi}_{qs} - \bar{\psi}_{mq}) \tag{3.49}$$

$$\bar{i}_{fd} = \frac{1}{\bar{L}_{lfd}}(\bar{\psi}_{fd} - \bar{\psi}_{md}) \tag{3.50}$$

$$\bar{i}_{kd} = \frac{1}{\bar{L}_{lkd}}(\bar{\psi}_{kd} - \bar{\psi}_{md}) \tag{3.51}$$

$$\bar{i}_{kq1} = \frac{1}{\bar{L}_{lkq1}}(\bar{\psi}_{kq1} - \bar{\psi}_{mq}) \tag{3.52}$$

$$\bar{i}_{kq2} = \frac{1}{\bar{L}_{lkq2}}(\bar{\psi}_{kq2} - \bar{\psi}_{mq}) \tag{3.53}$$

The non-reduced order model of the synchronous generator includes stator transients and rotor transients as well as the damper windings. The following differential equations are directly derived from Eqs (3.36), (3.37), (3.49) and (3.50):

$$\frac{d}{dt}\overline{\psi}_{ds} = \omega_b\left[\overline{v}_{ds} + \overline{\omega}_s\overline{\psi}_{qs} + \frac{\overline{r}_s}{\overline{L}_{ls}}(\overline{\psi}_{md} - \overline{\psi}_{ds})\right] \tag{3.54}$$

$$\frac{d}{dt}\overline{\psi}_{qs} = \omega_b\left[\overline{v}_{qs} - \overline{\omega}_s\overline{\psi}_{ds} + \frac{\overline{r}_s}{\overline{L}_{ls}}(\overline{\psi}_{mq} - \overline{\psi}_{qs})\right] \tag{3.55}$$

The rotor dynamic equations, with two damper windings in the q axis and one in the d axis, are as follows [from Eqs (3.40)–(3.43) and (3.50)–(3.53)]:

$$\frac{d}{dt}\overline{\psi}_{fd} = \omega_b\left[\frac{\overline{r}_{fd}}{\overline{L}_{md}}\overline{e}_{xfd} + \frac{\overline{r}_{fd}}{\overline{L}_{lfd}}(\overline{\psi}_{md} - \overline{\psi}_{fd})\right] \tag{3.56}$$

$$\frac{d}{dt}\overline{\psi}_{kd} = \omega_b\left[\overline{v}_{kd} + \frac{\overline{r}_{kd}}{\overline{L}_{lkd}}(\overline{\psi}_{md} - \overline{\psi}_{kd})\right] \tag{3.57}$$

$$\frac{d}{dt}\overline{\psi}_{kq1} = \omega_b\left[\overline{v}_{kq1} + \frac{\overline{r}_{kq1}}{\overline{L}_{lkq1}}(\overline{\psi}_{mq} - \overline{\psi}_{kq1})\right] \tag{3.58}$$

$$\frac{d}{dt}\overline{\psi}_{kq2} = \omega_b\left[\overline{v}_{kq2} + \frac{\overline{r}_{kq2}}{\overline{L}_{lkq2}}(\overline{\psi}_{mq} - \overline{\psi}_{kq2})\right] \tag{3.59}$$

The excitation dynamics of the generator are given by Eq. (3.56) where $\overline{e}_{xfd} = \overline{L}_{md} \cdot \overline{i}_{rfd}$ represents the excitation voltage of the generator at base speed ω_b.

If the zero sequence currents are present in the stator, then the following equation should also be considered:

$$\frac{d}{dt}\psi_{0s} = \omega_b\left(v_{0s} - \frac{\overline{r}_s}{\overline{L}_{ls}}\psi_{0s}\right) \tag{3.60}$$

3.8   Reduced-order Model

A reduced-order model may be obtained by neglecting the stator transients in Eqs (3.54), (3.55) and (3.60) as follows:

$$\overline{\psi}_{ds} = \frac{1}{\overline{\omega}_s}\left[\overline{v}_{qs} + \frac{\overline{r}_s}{\overline{L}_{ls}}(\overline{\psi}_{mq} - \overline{\psi}_{qs})\right] \tag{3.61}$$

$$\overline{\psi}_{qs} = -\frac{1}{\overline{\omega}_s}\left[\overline{v}_{ds} + \frac{\overline{r}_s}{\overline{L}_{ls}}(\overline{\psi}_{md} - \overline{\psi}_{ds})\right] \tag{3.62}$$

$$\overline{\psi}_{0s} = \frac{\overline{L}_{ls}}{\overline{r}_s}\overline{v}_{0s} \tag{3.63}$$

3.9   Control of Large Synchronous Generators

A large power system consists of a number of generators and loads connected through transmission and distribution circuits. Loads connected to the power system have different characteristics and vary continuously in time. In order to operate the power system within the limits required (voltage and frequency) and in order to maintain the stability of the system in case of a disturbance, large generators are controlled individually and collectively. The different controls associated with a synchronous generator are shown in Figure 3.10 (Kundur, 1994). These functional blocks perform two basic control actions, namely reactive power/voltage control and active power/frequency control.

3.9.1   Excitation Control

As conditions vary on the power system, the active and reactive power demand varies. Under heavy-load conditions, both the transmission system and the loads absorb reactive power and the synchronous generators need to inject reactive power into the network. Under light-load conditions, the capacitive behaviour of the transmission lines can become dominant and under such conditions it is desirable for synchronous generators to absorb reactive power. The variations in reactive power demand on a synchronous generator can be accommodated by adjusting its excitation voltage. The excitation system performs the basic function of automatic voltage regulation. It also performs the protective functions required to operate the machine and other equipment

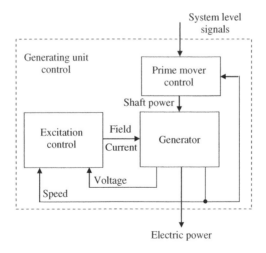

Figure 3.10   Synchronous generator control (Kundur, 1994)

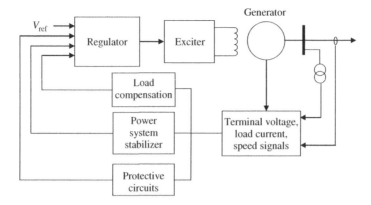

Figure 3.11   Block diagram of an excitation control system

within their capabilities. A block diagram of an excitation system is shown in Figure 3.11.

3.9.1.1   Regulator

A synchronous generator employs an automatic voltage regulator (AVR) to maintain the generator stator terminal voltage close to a predefined value. If the generator terminal voltage falls due to increased reactive power demand, the change in voltage is detected and a signal is fed into the exciter to produce an increase in excitation voltage. The generator reactive power output is thereby increased and the terminal voltage is returned close to its initial value.

3.9.1.2   Exciter

The purpose of an exciter is to supply an adjustable direct current to the main generator field winding. The exciter may be a DC generator on small set sizes. On larger sets, commutation problems prohibit the use of DC generators and AC generators are employed, supplying the field via a rectifier. Static excitation systems are also widely used. These comprise a controlled rectifier usually powered from the generator terminals and permit fast response excitation control. In all the mentioned cases, the DC supply is connected to the synchronous generator field via slip rings.

The necessity of employing slip rings and avoiding the associated maintenance requirement can be removed by employing brushless-excitation systems. Here the AC generator has its field mounted on the stator and its three-phase output on the rotor. This permits the rectifier to be mounted on the common

exciter-generator shaft and enables a direct connection to the generator field
to be made and thus avoids the necessity of employing slip rings.

3.9.1.3 Load Compensation

The AVR normally controls the generator stator terminal voltage. Building an
additional loop to the AVR control allows the voltage at a remote point on
the network to be controlled. The load compensator has adjustable resistance
and reactance that simulates the impedance between the generator terminals
and the point at which the voltage is being effectively controlled. Using this
impedance and the measured current, the voltage drop is computed and added
to the terminal voltage.

3.9.1.4 Power System Stabilizer

The basic function of the power system stabilizer (PSS) is to add damping
to the generator rotor oscillations by controlling its excitation. The commonly
used auxiliary stabilizing signals to control the excitation are shaft speed,
terminal frequency and power.

3.9.2 Prime Mover Control

The governing systems of the generator prime movers provide the means
of adjusting the power outputs of the generators of the network to match
the power demand of the network load. If, for example, the network load
increases, then this imposes increased torques on the generators, which causes
them to decelerate. The resulting decrease in speed is detected by the governor
of each regulating prime mover and used to increase its power output. The
change in power produced in an individual generator is determined by the
droop setting of its governor. A 4% droop setting indicates that the regulation
is such that a 4% change in speed would result in a 100% change in the
generator power output. At the steady state, all the generators of the network
operate at the same frequency and this frequency determines the operating
speeds of the individual generator prime movers. Hence, following a network
load increase, the network frequency will fall until the sum of the power output
changes that it produces in the regulating generators matches the change in
the network load. The basic elements of a governor power control loop are
shown in a block diagram in Figure 3.12 and the droop characteristic is shown
graphically in Figure 3.13. By changing the load reference set point, P_{ref}, the
generator governor characteristics (Figure 3.13) can be set to give the reference

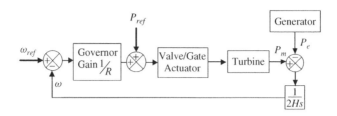

Figure 3.12 Speed governor system (Wood and Wollenberg, 1996)

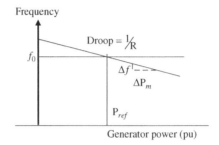

Figure 3.13 Droop characteristic

frequency, f_0 (50 or 60 Hz), at any desired unit output. In other words, it shifts the characteristic vertically.

References

Fitzgerald, A. E., Kingsley, C. Jr and Umans, S. D. (1992) *Electrical Machinery*, McGraw-Hill, New York.

Hindmarsh, J. and Renfrew, A. (1996) *Electrical Machines and Drive Systems*, Butterworth-Heinemann, Oxford.

Krause, P. C., Wasynczuk, O. and Shudhoff, S. D. (2002) *Analysis of Electric Machinery and Drive Systems*, 2nd edn, Wiley-IEEE Press, New York.

Kundur, P. (1994) *Power System Stability and Control*, McGraw-Hill, New York, ISBN 0-07-035958-X.

Wood, A. J. and Wollenberg, B. F. (1996) *Power Generation, Operation and Control*, 2nd edn, John Wiley & Sons, Inc., New York.

4

Fixed-speed Induction Generator (FSIG)-based Wind Turbines

4.1 Induction Machine Construction

Figure 4.1 shows a schematic diagram of the cross-section of a three-phase induction machine with one pair of field poles.

The stator consists of three-phase windings, as, bs and cs, distributed $120°$ apart in space. The rotor circuits have three distributed windings, ar, br and cr. The angle θ is given as the angle by which the axis of the phase ar rotor winding leads the axis of phase as stator winding in the direction of rotation and ω_r is the rotor angular velocity in electrical radians per second. The angular velocity of the stator field in electrical radians per second is represented by ω_s.

When balanced three-phase currents flow through the stator windings, a field rotating at synchronous speed, ω_s, is generated. The synchronous speed, ω_s (rad s^{-1}), is expressed as[1]

$$\omega_s = \frac{4\pi f_s}{p_f} \tag{4.1}$$

where f_s (Hz) is the frequency of the stator currents and p_f is the number of poles. If there is relative motion between the stator field and the rotor, voltages of frequency f_r (Hz) are induced in the rotor windings. The frequency f_r is equal to the slip frequency sf_s, where the slip, s, is given by

$$s = \frac{\omega_s - \omega_r}{\omega_s} \tag{4.2}$$

---

[1] Generally represented as $\frac{120 f_s}{p_f}$ in rev min^{-1}.

Wind Energy Generation: Modelling and Control   Olimpo Anaya-Lara, Nick Jenkins, Janaka Ekanayake, Phill Cartwright and Mike Hughes
© 2009 John Wiley & Sons, Ltd

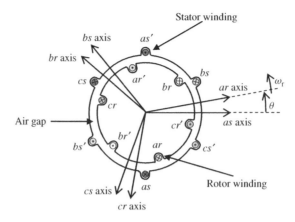

Figure 4.1   Schematic diagram of a three-phase induction machine (Kundur, 1994)

and ω_r (rad s^{-1}) is the rotor speed. The slip is positive if the rotor runs below the synchronous speed and negative if it runs above the synchronous speed (Kundur, 1994; Krause *et al.*, 2002).

The rotor of an induction machine may be one of two types: the squirrel-cage rotor and the wound rotor.

4.1.1   Squirrel-cage Rotor

The rotor of a squirrel-cage machine carries a winding consisting of a series set of bars in the rotor slots which are short-circuited by end rings at each end of the rotor. In use, the squirrel cage adopts the current pattern and pole distribution of the stator, enabling a basic rotor to be used for machines with differing pole numbers. However, for analysis purposes a squirrel-cage rotor may be treated as a symmetrical, short circuited star-connected three-phase winding.

4.1.2   Wound Rotor

The rotor of a wound-rotor machine carries a three-phase distributed winding with the same number of poles as the stator. This winding is usually star connected with the ends of the winding brought out to three slip-rings, enabling external circuits to be added to the rotor for control purposes.

4.2   Steady-state Characteristics

Figure 4.2 shows the steady-state, per-phase, equivalent circuit of the induction machine with all quantities in this circuit referred to the stator (using motor

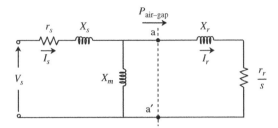

Figure 4.2 Single-phase equivalent circuit of an induction machine (Kundur, 1994; Fox *et al.*, 2007)

convention) (Fox *et al.*, 2007). In this figure $X_s = \omega_s L_s$ is the stator leakage reactance, $X_r = \omega_s L_r$ is the rotor leakage reactance and $X_m = \omega_s L_m$ is the magnetizing reactance. The terminal voltage, V_s, stator current, I_s, and rotor current, I_r, are per-phase RMS quantities.

The power transferred across the air-gap to the rotor (of one phase) is

$$P_{\text{air-gap}} = \frac{r_r}{s} I_r^2 \tag{4.3}$$

The torque developed by the machine (three-phase) is given by

$$T_e = 3 \frac{p_f}{2} \frac{r_r}{s\omega_s} I_r^2 \tag{4.4}$$

where $\omega_s = 2\pi f_s$. As can be seen in Eq. (4.4), the torque is slip dependent. For simple analysis of torque–slip relationships, the equivalent circuit of Figure 4.2 may be simplified by moving the magnetizing reactance to the terminals as shown in Figure 4.3. From this figure, the rotor current is

$$\mathbf{I_r} = \frac{\mathbf{V}_s}{\left(r_s + \dfrac{r_r}{s}\right) + j(X_s + X_r)} \tag{4.5}$$

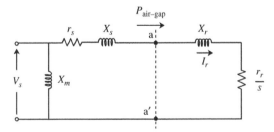

Figure 4.3 Equivalent circuit suitable for evaluating simple torque–slip relationships

Then from Eq. (4.4), the torque is

$$T_e = 3\frac{p_f}{2}\left(\frac{r_r}{s\omega_s}\right)\frac{V_s^2}{\left(r_s + \frac{r_r}{s}\right)^2 + (X_s + X_r)^2}$$

(4.6)

From Eq. (4.6) a typical relationship between torque and slip is shown in Figure 4.4. At standstill the speed is zero and the slip, s, is equal to 1 per unit (pu). Between zero and synchronous speed, the machine performs as a motor. Beyond synchronous speed, the machine performs as a generator.

Figure 4.5 shows the effect of varying the rotor resistance, r_r, on the torque of the induction machine (Fox et $al.$, 2007). A low rotor resistance is required to achieve high efficiency under normal operating conditions, but a high rotor resistance is required to produce high slip.

One way of controlling the rotor resistance (and therefore the slip and speed of the generator), is to use a wound rotor connected to external variable resistors through brushes and slip-rings. The rotor resistance is then adjusted by means of electronic equipment.

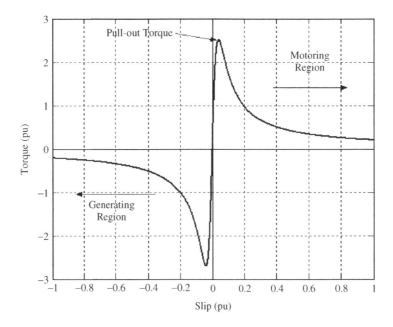

Figure 4.4 Typical torque–slip characteristic of an induction machine

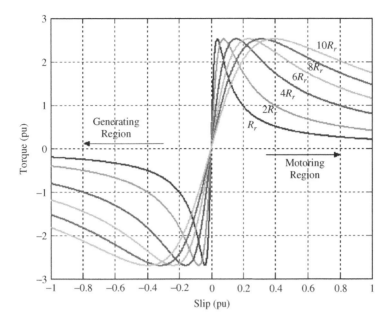

Figure 4.5   Torque–slip curves showing the effect of increased rotor circuit resistance (Fox *et al.*, 2007)

4.2.1   Variations in Generator Terminal Voltage

When an induction generator experiences a voltage sag on the network, the generator speed increases. Therefore, if the generator is not disconnected from the network in an appropriate time, it can accelerate to an unstable condition. Figure 4.6a and b show the steady-state torque–slip curves when the induction generator experiences a terminal voltage reduction (Fox *et al.*, 2007). Following a voltage drop, the machine moves to point Y, at which point the machine speeds up because the mechanical torque is higher than the electrical torque. It is then possible that the machine could move to point Z, which can lead to an unstable condition (point Z is beyond the 'pull-out' torque and the speed increases continually).

4.3   FSIG Configurations for Wind Generation

The typical configuration of a FSIG wind turbine using a squirrel-cage induction generator is shown in Figure 1.6 (Holdsworth *et al.*, 2003). In a squirrel-cage induction generator, the slip (and hence the rotor speed) vary

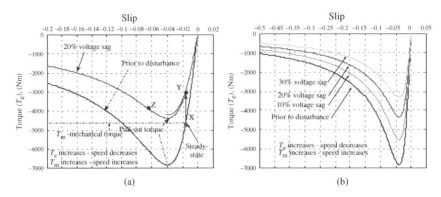

Figure 4.6 Induction machine torque–slip characteristics for variations in generator terminal voltage (Fox *et al.*, 2007)

with the amount of power generated. However, these rotor speed variations are very small (1–2%) and therefore it is normally referred to as constant-speed or fixed-speed wind turbine.

The generator typically operates at 690 V (line–line) and transmits power via vertical pendant cables to a switchboard and local transformer usually located in the tower base. As induction generators always consume reactive power, capacitor banks are employed to provide the reactive power consumption of the FSIG and improve the power factor. An anti-parallel thyristor soft-start unit is used to energize the generator once its operating speed is reached. The function of the soft-start unit (as described in Chapter 2) is to build up the magnetic flux slowly and hence minimize transient currents during energization of the generator (Fox *et al.*, 2007).

4.3.1 Two-speed Operation

Wind turbine rotors develop their peak efficiency at one particular tip-speed ratio. Energy capture can be increased by varying the rotational speed with the wind speed so that the turbine is always running at optimum tip-speed ratio. Alternatively, a slightly reduced improvement can be obtained by running the turbine at one of two fixed speeds so that the tip-speed ratio is closer to the optimum than with a single fixed speed (Burton *et al.*, 2001).

Two-speed operation is relatively expensive to implement if separate generators are used for each speed of turbine rotation. Either generators with differing numbers of poles may be connected to gearbox output shafts rotating at the same speed or generators with the same number of poles are connected to output shafts rotating at different speeds. The rating of the generator for low-speed operation would normally be much less than the turbine rating.

The development of induction generators with two sets of windings allows the number of poles within a single generator to be varied by connecting them together in different ways, a technique known as pole amplitude modulation (PAM) (Eastham and Balchin 1975; Rajaraman, 1977). Alternatively, two independent windings may be placed on the same stator. Generators of this type are available which can be switched between four- and six-pole operation, giving a speed ratio of 1.5. With the correct selection of the two operating speeds, this ratio produces a significant increase in energy capture and reduces the rotor tip speed at low wind speeds, when environmental noise constraints are most onerous.

4.3.2 Variable-slip Operation

The variable-slip generator is essentially an induction generator with a variable resistor in series with the rotor circuit, controlled by a high-frequency semiconductor switch as shown in Figure 4.7. Below rated wind speed and power, this acts just like a conventional fixed-speed induction generator. Above rated, however, control of the resistance effectively allows the air-gap torque to be controlled and the slip speed to vary, so that behaviour is then similar to that of a variable-speed system. A speed range of about 10% is typical with a consequent energy loss of 10% in the additional resistor (Burton *et al.*, 2001).

Slip-rings can be avoided by mounting the variable resistors and control circuitry on the generator rotor. An advantage of mounting these externally via slip-rings is that it is then easier to dissipate the extra heat which is generated above rated and which may otherwise be a limiting factor at large sizes. In some configurations, all control functions are executed by means of fibre-optic circuits (Thiringer *et al.*, 2003).

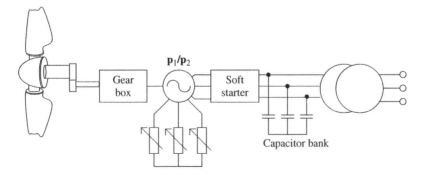

Figure 4.7 Configuration for variable-slip operation

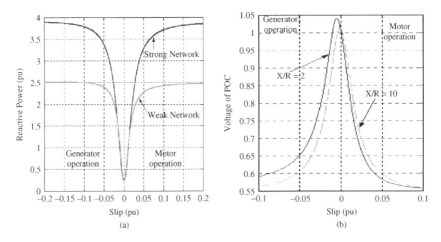

Figure 4.8 Reactive power drawn by an induction generator and voltage at the point of connection. (a) Reactive power variation with slip; (b) voltage at the point of connection

4.3.3 Reactive Power Compensation Equipment

The induction machine needs reactive power to build up the magnetic field. It is known that reactive power does not contribute to direct energy conversion. The current associated with it, the reactive current, however, causes losses in the supply and in the machine. The higher the reactive current content in the overall current, the lower is the power factor $\cos \phi$. Since the induction machine is not 'excited' like the synchronous machine, it takes the reactive power from the grid.

Figure 4.8 shows the variation of the reactive power absorbed by the induction generator with slip. Two cases are considered, where the induction machine is connected to a strong network (fault level of 3600 MVA) and to a weak network (fault level of 360 MVA) through a transmission line having an X/R ratio of 10. It can be seen from the figure that as the slip or the power generation increases, the amount of reactive power absorbed by the generator also increases. Due to the large amount of reactive power drawn from the network, the voltage across the transmission line drops. The voltage at the point of connection with the network decreases as the slip increases.

4.4 Induction Machine Modelling

As shown in Figure 4.1, the stator of the induction machine carries three-phase windings and is connected to a three-phase voltage source. The windings produce a magnetic field rotating at synchronous speed. As in the case of the synchronous machine, this equivalent rotating magnetic field can be represented by two coils, one on the d axis and the other on the q axis, which rotate at the synchronous speed of the supply voltage. Once the stator magnetic field cuts the rotor conductors, three-phase voltages of slip frequency sf_s are induced on the rotor. As the rotor conductors are normally short-circuited, three-phase currents at slip frequency flow on the rotor. These currents also produce a rotating magnetic field which is rotating at slip speed ($\omega_s - \omega_r = s\omega_s$) with respect to the rotor. A viewer standing on the Earth sees that the rotor magnetic field also rotates at the synchronous speed ($s\omega_s + \omega_r = \omega_s$). Therefore, the rotor magnetic field can also be represented by two perpendicular coils with respect to the stator d and q axes, which are rotating at the synchronous speed of the supply. In the synchronous reference frame, all the coils are then stationary and thus inductances are constant. Now the machine voltage and flux equations can be expressed in dq components as shown in Table 4.1.

4.4.1 FSIG Model as a Voltage Behind a Transient Reactance

A conventional modelling technique in power systems is to represent the FSIG by a simple voltage behind a transient reactance equivalent circuit. The stator voltage is expressed in terms of a voltage behind a transient reactance by substituting the stator fluxes $\overline{\psi}_{ds}$ and $\overline{\psi}_{qs}$ [Eqs (4.11) and (4.12)] in the stator voltage equations Eqs (4.7) and (4.8) as follows:

$$\overline{v}_{ds} = -\overline{r}_s\overline{i}_{ds} - \overline{\omega}_s(-\overline{L}_{ss}\overline{i}_{qs} + \overline{L}_m\overline{i}_{qr}) + \frac{1}{\omega_b}\frac{d}{dt}(-\overline{L}_{ss}\overline{i}_{ds} + \overline{L}_m\overline{i}_{dr}) \quad (4.15)$$

$$\overline{v}_{qs} = -\overline{r}_s\overline{i}_{qs} + \overline{\omega}_s(-\overline{L}_{ss}\overline{i}_{ds} + \overline{L}_m\overline{i}_{dr}) + \frac{1}{\omega_b}\frac{d}{dt}(-\overline{L}_{ss}\overline{i}_{qs} + \overline{L}_m\overline{i}_{qr}) \quad (4.16)$$

Table 4.1 Induction machine equations in dq coordinates (in per unit)

Voltage equations:		Flux equations:	
$\overline{v}_{ds} = -\overline{r}_s\overline{i}_{ds} - \overline{\omega}_s\overline{\psi}_{qs} + \frac{1}{\omega_b}\frac{d}{dt}\overline{\psi}_{ds}$	(4.7)	$\overline{\psi}_{ds} = -\overline{L}_{ss}\overline{i}_{ds} + \overline{L}_m\overline{i}_{dr}$	(4.11)
$\overline{v}_{qs} = -\overline{r}_s\overline{i}_{qs} + \overline{\omega}_s\overline{\psi}_{ds} + \frac{1}{\omega_b}\frac{d}{dt}\overline{\psi}_{qs}$	(4.8)	$\overline{\psi}_{qs} = -\overline{L}_{ss}\overline{i}_{qs} + \overline{L}_m\overline{i}_{qr}$	(4.12)
$\overline{v}_{dr} = \overline{r}_r\overline{i}_{dr} - s\overline{\omega}_s\overline{\psi}_{qr} + \frac{1}{\omega_b}\frac{d}{dt}\overline{\psi}_{dr}$	(4.9)	$\overline{\psi}_{dr} = \overline{L}_{rr}\overline{i}_{dr} - \overline{L}_m\overline{i}_{ds}$	(4.13)
$\overline{v}_{qr} = \overline{r}_r\overline{i}_{qr} + s\overline{\omega}_s\overline{\psi}_{dr} + \frac{1}{\omega_b}\frac{d}{dt}\overline{\psi}_{qr}$	(4.10)	$\overline{\psi}_{qr} = \overline{L}_{rr}\overline{i}_{qr} - \overline{L}_m\overline{i}_{qs}$	(4.14)

From the rotor flux equations Eqs (4.13) and (4.14), expressions are derived for the rotor currents, $\bar{i}_{dr}$ and $\bar{i}_{qr}$ and substituted in Eqs (4.15) and (4.16). Thus, the dq components of the stator voltage are expressed as a function of the voltage behind a transient reactance by

$$\bar{v}_{ds} = -\bar{r}_s\bar{i}_{ds} + \overline{X}'\bar{i}_{qs} + \bar{e}_d - \frac{\overline{X}'}{\omega_s}\frac{d}{dt}\bar{i}_{ds} + \frac{1}{\omega_s}\frac{d}{dt}\bar{e}_q \qquad (4.17)^2$$

$$\bar{v}_{qs} = -\bar{r}_s\bar{i}_{qs} - \overline{X}'\bar{i}_{ds} + \bar{e}_q - \frac{\overline{X}'}{\omega_s}\frac{d}{dt}\bar{i}_{qs} - \frac{1}{\omega_s}\frac{d}{dt}\bar{e}_d \qquad (4.18)^2$$

where $\bar{e}_d$ and $\bar{e}_q$ are the dq components of the voltage behind a transient reactance defined as

$$\bar{e}_d = -\frac{\omega_s\overline{L}_m}{\overline{L}_{rr}}\overline{\psi}_{qr} \qquad (4.19)$$

$$\bar{e}_q = \frac{\omega_s\overline{L}_m}{\overline{L}_{rr}}\overline{\psi}_{dr} \qquad (4.20)$$

and $\overline{X}'$ is the transient or short-circuit reactance of the induction machine:

$$\overline{X}' = \omega_s\left(\overline{L}_{ss} - \frac{\overline{L}_m^2}{\overline{L}_{rr}}\right) \qquad (4.21)$$

The stator currents as a function of the voltage behind a transient reactance can be derived directly from the stator voltages given by Eqs (4.17) and (4.18) as follows:

$$\frac{d}{dt}\bar{i}_{ds} = \frac{\omega_s}{\overline{X}'}\left(-\bar{r}_s\bar{i}_{ds} + \overline{X}'\bar{i}_{qs} + \bar{e}_d - \bar{v}_{ds} + \frac{1}{\omega_s}\frac{d}{dt}\bar{e}_q\right) \qquad (4.22)$$

$$\frac{d}{dt}\bar{i}_{qs} = \frac{\omega_s}{\overline{X}'}\left(-\bar{r}_s\bar{i}_{qs} - \overline{X}'\bar{i}_{ds} + \bar{e}_q - \bar{v}_{qs} - \frac{1}{\omega_s}\frac{d}{dt}\bar{e}_d\right) \qquad (4.23)$$

The equation of the voltage behind a transient reactance is derived from the rotor voltage and flux equations. From the rotor flux equations Eqs (4.13) and (4.14), expressions are obtained for the rotor currents, $\bar{i}_{dr}$ and $\bar{i}_{qr}$, and substituted in the rotor voltage equations Eqs (4.9) and (4.10):

$$\bar{v}_{dr} = \bar{r}_r\left(\frac{\overline{\psi}_{dr} + \overline{L}_m\bar{i}_{ds}}{\overline{L}_{rr}}\right) - \bar{s}\,\bar{\omega}_s\overline{\psi}_{qr} + \frac{1}{\omega_b}\frac{d}{dt}\overline{\psi}_{dr} \qquad (4.24)$$

---

^{2} ω_s in rad s^{-1}.

$$\overline{v}_{qr} = \overline{r}_r \left(\frac{\overline{\psi}_{qr} + \overline{L}_m \overline{i}_{qs}}{\overline{L}_{rr}} \right) + \overline{s} \, \overline{\omega}_s \overline{\psi}_{dr} + \frac{1}{\omega_b} \frac{d}{dt} \overline{\psi}_{qr} \qquad (4.25)$$

From Eqs (4.19) and (4.20), expressions are derived for the rotor fluxes, $\overline{\psi}_{dr}$ and $\overline{\psi}_{qr}$, in terms of $\overline{e}_d$ and $\overline{e}_q$, respectively, and substituted in Eqs (4.24) and (4.25) to obtain the derivatives of the voltage behind a transient reactance:

$$\frac{d\overline{e}_d}{dt} = -\frac{\omega_b}{\overline{T}_0}[\overline{e}_d - (\overline{X} - \overline{X}')\overline{i}_{qs}] + s\omega_s \overline{e}_q - \omega_s \frac{\overline{L}_m}{\overline{L}_{rr}} \overline{v}_{qr} \qquad (4.26)$$

$$\frac{d\overline{e}_q}{dt} = -\frac{\omega_b}{\overline{T}_0}[\overline{e}_q + (\overline{X} - \overline{X}')\overline{i}_{ds}] - \overline{s}\omega_s \overline{e}_d + \omega_s \frac{\overline{L}_m}{\overline{L}_{rr}} \overline{v}_{dr} \qquad (4.27)$$

where

$$\overline{T}_0 = \frac{\overline{L}_{rr}}{\overline{r}_r} = \frac{\overline{L}_r + \overline{L}_m}{\overline{r}_r} \qquad (4.28)$$

and

$$\overline{X} = \overline{\omega}_s \overline{L}_{ss} \qquad (4.29)$$

For a fixed-speed induction generator, $\overline{v}_{dr} = \overline{v}_{qr} = 0$.

The per unit reactance $\overline{X}$ is defined as the open-circuit reactance. The constant $\overline{T}_0$ is the per unit transient open-circuit time constant of the induction machine and in this form it is expressed in radians. The reduced order equations presented above are also applicable with time t and the time constant $\overline{T}_0$ expressed in seconds. The rotor current equations are derived by rearranging the expressions for $\overline{e}_d$ and $\overline{e}_q$ in terms of the rotor fluxes $\overline{\psi}_{dr}$ and $\overline{\psi}_{qr}$ and substituting those in Eqs (4.13) and (4.14) as follows:

$$\overline{i}_{dr} = \frac{\overline{\psi}_{dr} - \overline{L}_m \overline{i}_{ds}}{\overline{L}_{rr}} = \frac{1}{\overline{\omega}_s \overline{L}_m} \overline{e}_q + \frac{\overline{L}_m}{\overline{L}_{rr}} \overline{i}_{ds} \qquad (4.30)$$

$$\overline{i}_{qr} = \frac{\overline{\psi}_{qr} + \overline{L}_m \overline{i}_{qs}}{\overline{L}_{rr}} = -\frac{1}{\overline{\omega}_s \overline{L}_m} \overline{e}_d + \frac{\overline{L}_m}{\overline{L}_{rr}} \overline{i}_{qs} \qquad (4.31)$$

4.4.1.1 Rotor Mechanics Equation

To complete the FSIG dynamic model, the differential equations describing the electric voltage and current components of the machine need to be combined with a rotor mechanics equation. This equation is of major importance in power system stability analysis as it describes the effect of any mismatch between the electromagnetic torque and the mechanical torque of the machine. The rotor mechanics equation of the FSIG is given as

$$J \frac{\mathrm{d}}{\mathrm{d}t} \omega_r = T_m - T_e \tag{4.32}$$

where T_m (Nm) is the mechanical torque, T_e (Nm) is the electromagnetic torque and J (kgm^2) is the combined moment of inertia of generator and turbine.

Equation (4.32) can be normalized in terms of the per unit inertia constant, H, defined as the kinetic energy in watts seconds at rated speed divided by S_{base}. Using ω_s to denote rated angular velocity in mechanical radians per second, the inertia constant is (Kundur, 1994)

$$H = \frac{J \omega_s^2}{2 S_{base}} \tag{4.33}$$

The moment of inertia J in terms of H is

$$J = \frac{2H}{\omega_s^2} S_{base} \tag{4.34}$$

Substituting Eq. (4.34) in Eq. (4.32) and rearranging terms yields

$$\frac{2H}{\omega_s^2} S_{base} \frac{\mathrm{d}}{\mathrm{d}t} \omega_r = T_m - T_e \tag{4.35}$$

$$2H \frac{\mathrm{d}}{\mathrm{d}t} \left(\frac{\omega_r}{\omega_s} \right) = \frac{T_m - T_e}{S_{base}/\omega_s} \tag{4.36}$$

Considering that $T_{base} = S_{base}/\omega_s$ the rotor mechanics equation in per unit notation is

$$\frac{\mathrm{d}}{\mathrm{d}t} \overline{\omega}_r = \frac{1}{2H} (\overline{T}_m - \overline{T}_e) \tag{4.37}^3$$

where the electromagnetic torque, T_e, in per unit is calculated as

$$\overline{T}_e = \frac{\overline{e}_d \overline{i}_{ds} + \overline{e}_q \overline{i}_{qs}}{\overline{\omega}_s} \tag{4.38}$$

The complete fifth-order model of the FSIG is summarized in Table 4.2.

---

[3] It is worth noting that in pu power and torque are equal and therefore Eq. (4.37) can also be represented as Eq. (8.6) in Chapter 8.

Table 4.2 FSIG fifth-order model

Stator voltage:

$$\bar{v}_{ds} = -\bar{r}_s \bar{i}_{ds} + \overline{X}'\,\bar{i}_{qs} + \bar{e}_d - \frac{\overline{X}'}{\omega_s}\frac{d}{dt}\bar{i}_{ds} + \frac{1}{\omega_s}\frac{d}{dt}\bar{e}_q$$

$$\bar{v}_{qs} = -\bar{r}_s \bar{i}_{qs} - \overline{X}'\,\bar{i}_{ds} + \bar{e}_q - \frac{\overline{X}'}{\omega_s}\frac{d}{dt}\bar{i}_{qs} - \frac{1}{\omega_s}\frac{d}{dt}\bar{e}_d$$

Voltage behind a transient reactance:

$$\frac{d}{dt}\bar{e}_d = -\frac{1}{T_0}[\bar{e}_d - (\overline{X} - \overline{X}')\bar{i}_{qs}] + s\omega_s \bar{e}_q$$

$$\frac{d}{dt}\bar{e}_q = -\frac{1}{T_0}[\bar{e}_q + (\overline{X} - \overline{X}')\bar{i}_{ds}] - s\omega_s \bar{e}_d$$

Stator current:

$$\frac{d}{dt}\bar{i}_{ds} = \frac{\omega_s}{\overline{X}'}\left(-\bar{r}_s \bar{i}_{ds} + \overline{X}'\,\bar{i}_{qs} + \bar{e}_d - \bar{v}_{ds} + \frac{1}{\omega_s}\frac{d}{dt}\bar{e}_q\right)$$

$$\frac{d}{dt}\bar{i}_{qs} = \frac{\omega_s}{\overline{X}'}\left(-\bar{r}_s \bar{i}_{qs} - \overline{X}'\,\bar{i}_{ds} + \bar{e}_q - \bar{v}_{qs} - \frac{1}{\omega_s}\frac{d}{dt}\bar{e}_d\right)$$

Rotor mechanics equation:

$$\frac{d}{dt}\bar{\omega}_r = \frac{1}{2H}(\bar{T}_m - \bar{T}_e)$$

$$\bar{T}_e = \frac{\bar{e}_d \bar{i}_{ds} + \bar{e}_q \bar{i}_{qs}}{\bar{\omega}_s}$$

For representation of the FSIG in power system stability studies, it is an accepted practice to reduce the mathematical model to a third-order form (Tande, 2003). The differential terms representing the stator transients are then neglected. Neglecting these corresponds to ignoring the DC component in the stator transient current. This simplification is useful for large system modelling to ensure compatibility with the models representing other system components, particularly the transmission network. The simplification to derive the third-order model of the FSIG is achieved straightforwardly from the stator voltage equation given by Eqs (4.17) and (4.18). After neglecting the differential terms in these equations, we obtain

$$\bar{v}_{ds} = -\bar{r}_s \bar{i}_{ds} + \overline{X}' \bar{i}_{qs} + \bar{e}_d \qquad (4.39)$$

$$\bar{v}_{qs} = -\bar{r}_s \bar{i}_{qs} - \overline{X}' \bar{i}_{ds} + \bar{e}_q \qquad (4.40)$$

With the stator transients neglected, the FSIG third-order model can be summarized as shown in Table 4.3.

4.5 Dynamic Performance of FSIG Wind Turbines

4.5.1 Small Disturbances

The dynamic performance of an FSIG wind turbine is illustrated using the two-machine network shown in Figure 4.9. In this network the FSIG is connected to an infinite bus through the impedances of the turbine transformer

Table 4.3 FSIG third-order model

Stator voltage:	Stator current:
$\bar{v}_{ds} = -\bar{r}_s \bar{i}_{ds} + \overline{X}' \bar{i}_{qs} + \bar{e}_d$	$\bar{i}_{ds} = \dfrac{1}{\left(\bar{r}_s^2 + \overline{X}'^2\right)}[(\bar{e}_d - \bar{v}_{ds})\bar{r}_s + (\bar{e}_q - \bar{v}_{qs})\overline{X}']$
$\bar{v}_{qs} = -\bar{r}_s \bar{i}_{qs} - \overline{X}' \bar{i}_{ds} + \bar{e}_q$	$\bar{i}_{qs} = \dfrac{1}{\left(\bar{r}_s^2 + \overline{X}'^2\right)}[(\bar{e}_q - \bar{v}_{qs})\bar{r}_s - (\bar{e}_d - \bar{v}_{ds})\overline{X}']$
Voltage behind a transient reactance:	**Rotor mechanics equation:**
$\dfrac{d}{dt}\bar{e}_d = -\dfrac{1}{T_0}[\bar{e}_d - (\overline{X} - \overline{X}')\bar{i}_{qs}] + s\omega_s\bar{e}_q$	$\dfrac{d}{dt}\bar{\omega}_r = \dfrac{1}{2H}(\overline{T}_m - \overline{T}_e)$
$\dfrac{d}{dt}\bar{e}_q = -\dfrac{1}{T_0}[\bar{e}_q + (\overline{X} - \overline{X}')\bar{i}_{ds}] - s\omega_s\bar{e}_d$	$\overline{T}_e = \dfrac{\bar{e}_d\bar{i}_{ds} + \bar{e}_q\bar{i}_{qs}}{\bar{\omega}_s}$

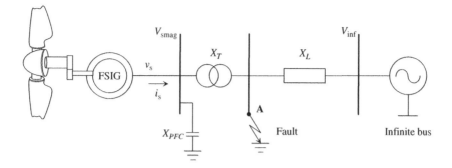

Figure 4.9   Connection of the FSIG-based wind turbine to an infinite bus

and a single transmission line. Capacitive compensation is provided on the generator terminals.

4.5.1.1   Step Change in Mechanical Torque Input

To illustrate the performance of the FSIG wind turbine in this situation, a decrease of 20% in the mechanical input torque, T_m, is applied at $t = 1$ s. Figure 4.10 shows the terminal voltage, V_{smag}, electrical torque, T_e, and slip,

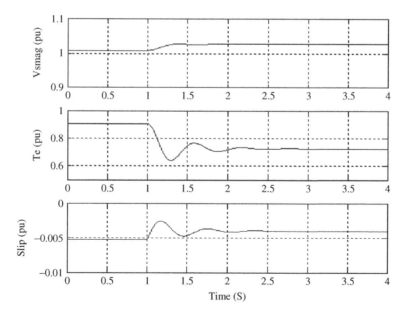

Figure 4.10   FSIG responses for a 20% decrease in the mechanical torque input T_m at $t = 1$ s

s, of the FSIG. The electrical torque output of the FSIG follows the new torque reference after a short transient period. It is seen that the speed of the FSIG also decreases as the mechanical input torque decreases. However, as the FSIG is provided with no mechanical or electrical control, it is observed that the variations in the input torque influence the profile of the terminal voltage, V_{smag}.

4.5.1.2 Step Change in the Infinite Bus Voltage

A variation in the voltage of the network to which the FSIG wind turbine is connected has a significant impact on the performance of the generator and even transient failures may be encountered due to system voltage collapse, which causes induction generators runaway. Two cases are illustrated where the voltage of the infinite bus, V_{inf}, is reduced first 20% and then 60% below the nominal operating level. In both cases the voltage dip is sustained for a period of 500 ms.

The FSIG responses when V_{inf} is reduced by 20% are shown in Figure 4.11. As V_{inf} drops the electrical torque of the generator, T_e, decreases but recovers to its initial value. However, the generator continually accelerates while the voltage is low. When the infinite bus voltage is re-established to the nominal value after 500 ms, the system recovers the initial operating conditions after

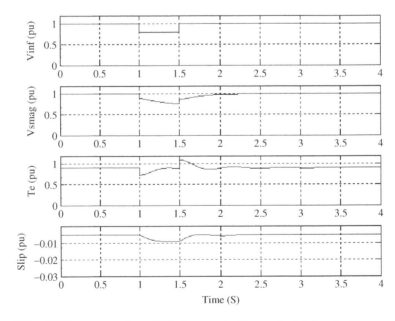

Figure 4.11 FSIG responses for a 20% decrease in V_{inf} applied at $t = 1$ s for a period of 500 ms

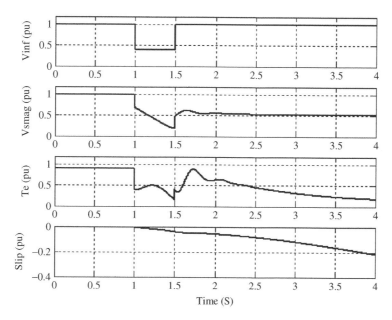

Figure 4.12 FSIG responses for a 60% decrease in V_{inf} applied at $t = 1$ s for a period of 500 ms

a short transient period has elapsed. Although the generator's torque recovers normal operation during the voltage dip, the generator will eventually runway if the voltage dip is sustained for a longer period of time.

Figure 4.12 shows the FSIG responses for a 60% decrease in V_{inf}. In this situation, the generator is unable to continue to operate and is unstable even when the infinite bus voltage is re-established to the nominal value after 500 ms. As in this case the voltage drop at the generator terminals is significant, the rotor continues to accelerate, in which case the reactive power consumption is higher and therefore the generator voltage fails to recover.

4.5.2 Performance During Network Faults

To illustrate the response of the FSIG wind turbine to large power system disturbances, the simple network model of Figure 4.9 is used. A three-phase balanced fault is applied at $t = 1$ s at the high-voltage terminals of the FSIG transformer (Point A). The FSIG wind turbine is studied using both third- and fifth-order models. Figure 4.13 shows the FSIG responses (terminal voltage, V_{smag}, and electrical torque, T_e) obtained with both third- and fifth-order models with a fault clearance time of 140 ms. During the fault the generator overspeeds and the terminal voltage starts to reduce further as more reactive

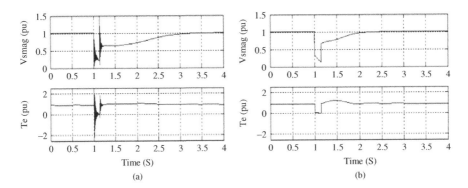

Figure 4.13 FSIG performance during faults. Fault applied at $t = 1$ s and cleared after 140 ms. (a) Fifth-order model; (b) third-order model

power is being absorbed by the generator. However, when the fault is cleared the generator and the network recover stability.

The responses in Figure 4.13 also illustrate the significance of neglecting the stator transients in the reduced third-order model. When the stator transients are neglected, the FSIG responses contain only the fundamental frequency component. Figure 4.14 shows the dq components of the stator current where the oscillations due to the stator transients are observed in the fifth-order model responses (Figure 4.14a). These oscillations are also present in the responses of the terminal voltage and torque of the generator as shown in Figure 4.15a.

A critical factor that limits the operation of an FSIG is the maximum fault clearance time that the generator can withstand before going into instability

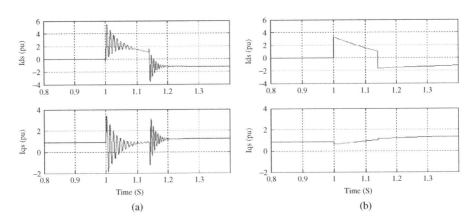

Figure 4.14 FSIG performance during faults: dq components of the stator currents during a fault applied at $t = 1$ s and cleared after 140 ms. (a) Fifth-order model; (b) third-order model

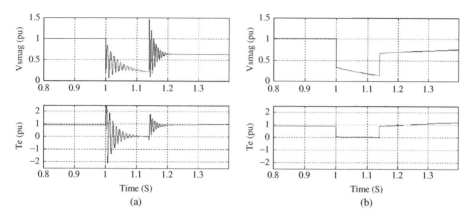

Figure 4.15 FSIG performance during faults. Fault applied at $t = 1$ s and cleared after 140 ms. (a) Fifth-order model; (b) third-order model

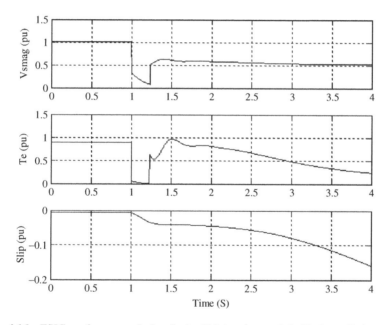

Figure 4.16 FSIG performance during faults (third-order model). Fault applied at $t = 1$ s and cleared after 230 ms. The system is unstable for a fault that lasts longer than 220 ms

(runaway). For the specific FSIG and network parameters used in this example, the maximum fault clearance time is around 220 ms. If the fault remains longer, the system loses stability, as illustrated in Figure 4.16.

References

Burton, T., Sharpe, D., Jenkins, N. and Bossanyi, E. (2001) *Wind Energy Handbook*, John Wiley & Sons, Ltd, Chichester, ISBN 10: 01471489972.

Eastham, J. F. and Balchin, M. J. (1975) Pole-change windings for linear induction motors, *Proceedings of the IEEE*, **122** (2), 154–160.

Fox, B., Flynn, D., Bryans, L., Jenkins, N., Milborrow, D., O'Malley, M., Watson, R. and Anaya-Lara, O. (2007) *Wind Power Integration: Connect and System Operational Aspects*, IET Power and Energy Series, Vol. 50, Institution of Engineering and Technology, Stevenage, ISBN 10: 0863414494.

Holdsworth, L., Wu, X., Ekanayake, J. B. and Jenkins, N. (2003) Comparison of fixed speed and doubly-fed induction wind turbines during power system disturbances, *IEE Proceedings Generation, Transmission and Distribution* **150** (3), 343–352.

Krause, P. C., Wasynczuk, O. and Shudhoff, S. D. (2002) *Analysis of Electric Machinery and Drive System*, 2nd edn, Wiley-IEEE Press, New York.

Kundur, P. (1994) *Power System Stability and Control*, McGraw-Hill, New York, ISBN 0-07-035958-X.

Rajaraman, K. C. (1977) Design criteria for pole-changing windings, *Proceedings of the IEE*, **124** (9), 775–783.

Tande, J. O. (2003) Grid integration of wind farms, *Wind Energy Journal*, **6**, 281–295.

Thiringer, T., Petersson, A. and Petru, T. (2003) Grid disturbance response of wind turbines equipped with induction generator and doubly-fed induction generator, *Power Engineering Society General Meeting*, IEEE, Vol. 3, pp. 13–17.

5

Doubly Fed Induction Generator (DFIG)-based Wind Turbines

5.1 Typical DFIG Configuration

A typical configuration of a DFIG wind turbine is shown in Figure 1.7. It uses a wound-rotor induction generator with slip-rings to transmit current between the converter and the rotor windings and variable-speed operation is obtained by injecting a controllable voltage into the rotor at the desired slip frequency (Holdsworth *et al.*, 2003). The rotor winding is fed through a variable-frequency power converter, typically based on two AC/DC IGBT-based voltage source converters (VSCs), linked through a DC bus. The variable-frequency rotor supply from the converter enables the rotor mechanical speed to be decoupled from the synchronous frequency of the electrical network, thereby allowing variable-speed operation of the wind turbine. The generator and converters are protected by voltage limits and an over-current 'crowbar'.

A DFIG wind turbine can transmit power to the network through both the generator stator and the converters. When the generator operates in super-synchronous mode, power will be delivered from the rotor through the converters to the network, and when the generator operates in sub-synchronous mode, the rotor will absorb power from the network through the converters. These two modes of operation are illustrated in Figure 5.1, where ω_s is the synchronous speed of the stator field and ω_r is the rotor speed (Fox *et al.*, 2007).

5.2 Steady-state Characteristics

The steady-state performance can be described using the Steinmetz per phase equivalent circuit model shown in Figure 5.2, where motor convention is used

Wind Energy Generation: Modelling and Control   Olimpo Anaya-Lara, Nick Jenkins,
Janaka Ekanayake, Phill Cartwright and Mike Hughes
© 2009 John Wiley & Sons, Ltd

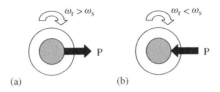

Figure 5.1 (a) Super-synchronous and (b) sub-synchronous operation of the DFIG wind turbine (Fox *et al.*, 2007)

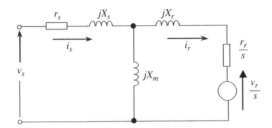

Figure 5.2 DFIG equivalent circuit with injected rotor voltage

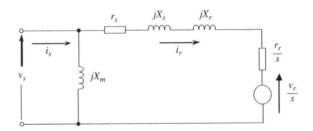

Figure 5.3 DFIG equivalent circuit with injected rotor voltage

(Hindmarsh, 1995). In this figure, v_s and v_r are the stator and rotor voltages, i_s and i_r are the stator and rotor currents, r_s and r_r are the stator and rotor resistances (per phase), X_s and X_r are the stator and rotor leakage reactances, X_m is the magnetizing reactance and s is the slip.

The equivalent circuit of Figure 5.2 can be simplified by transferring the magnetising branch to the terminals, as shown in Figure 5.3.

The torque–slip curves for the DFIG can be calculated from the approximate equivalent circuit model using the following equations (Hindmarsh, 1995). The rotor current can be calculated from

$$I_r = \frac{V_s - \left(\frac{V_r}{s}\right)}{\left(r_s + \frac{r_r}{s}\right) + j(X_s + X_r)} \tag{5.1}$$

The electrical torque, T_e, of the machine, which equates to the power balance across the stator to rotor gap, can be calculated from

$$T_e = \left(I_r^2 \frac{r_r}{s}\right) + \frac{P_r}{s} \tag{5.2}$$

where the power supplied or absorbed by the controllable-source injecting voltage into the rotor circuit, that is, the rotor active power, P_r, can be calculated from

$$P_r = \frac{V_r}{s} I_r \cos\theta; \quad P_r = \mathrm{Re}\left(\frac{V_r}{s} I_r^*\right) \tag{5.3}$$

Figure 5.4 shows the torque–slip characteristics of the DFIG with in-phase (V_{qr}) and out-of-phase (V_{dr}) components of the rotor voltage with respect to the stator voltage.

An example where rotor injection is used to drive the machine into sub- and super-synchronous speeds is given in Figure 5.5. With a negative injected voltage v_r, the speed of the machine will increase to super-synchronous operation as shown by curve 1. To reduce the speed of the machine to sub-synchronous operation, a positive voltage v_r is applied; the torque–slip characteristic for this case is illustrated by curve 2.

The DFIG operating in super synchronous speed (point A in Figure 5.5) will deliver power from the rotor through the converters to the network. At sub synchronous speed (point B in Figure 5.5) the DFIG rotor absorbs active power through the converters.

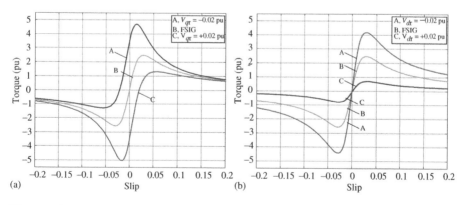

Figure 5.4 Torque–slip characteristic for the FSIG and DFIG. (a) With in-phase rotor injection; (b) with quadrature rotor injection

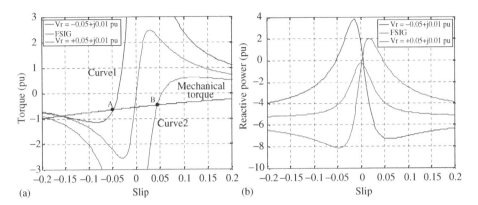

Figure 5.5   (a) Torque–slip and (b) reactive power–slip characteristic for the FSIG and DFIG

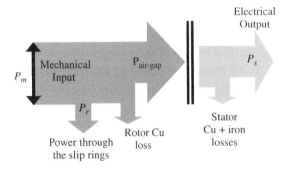

Figure 5.6   DFIG power relationships (Fox *et al.*, 2007)

5.2.1   Active Power Relationships in the Steady State

Figure 5.6 shows the steady-state relationship between mechanical power and rotor and stator electrical active powers in a DFIG system (Fox, *et al.*, 2007). In this figure, P_m is the mechanical power delivered by the turbine, P_r is the power delivered by the rotor to the converter, $P_{air\text{-}gap}$ is the power at the generator air-gap, P_s is the power delivered by the stator and P_g is the total power generated (by the stator plus the converter) and delivered to the grid. If the stator losses are neglected, then

$$P_{air\text{-}gap} = P_s \tag{5.4}$$

and neglecting rotor losses

$$P_{air\text{-}gap} = P_m - P_r \tag{5.5}$$

Combining Eqs (5.4) and (5.5), the stator power, P_s, can be expressed as

$$P_s = P_m - P_r \tag{5.6}$$

Equation (5.6) can be expressed in terms of the generator torque, T, as

$$T\omega_s = T\omega_r - P_r \tag{5.7}$$

where $P_s = T\omega_s$ and $P_m = T\omega_r$. Rearranging terms in Eq (5.7):

$$P_r = -T(\omega_s - \omega_r) \tag{5.8}$$

Then the stator and rotor powers can be related through the slip s as

$$P_r = -sT\omega_s = -sP_s \tag{5.9}$$

Combining Eqs (5.6) and (5.9), the mechanical power, P_m, can be expressed as

$$\begin{aligned} P_m &= P_s + P_r \\ &= P_s - sP_s \\ &= (1-s)P_s \end{aligned} \tag{5.10}$$

and the total power delivered to the grid, P_g, is then given by

$$P_g = P_s + P_r \tag{5.11}$$

The controllable range of s determines the size of the converters for the DFIG. Mechanical and other restrictions limit the maximum slip and a practical speed range may be between 0.7 and 1.2 pu.

5.2.2 Vector Diagram of Operating Conditions

Figure 5.7 shows the vector diagram of operating conditions for the DFIG (operating in super synchronous mode) (Anaya-Lara et al., 2006). This vector diagram provides an understanding of the way in which the machine is controlled and it can be readily employed for control design purposes. The bold font notation is used to represent a vector. $\boldsymbol{E}_g$ is the voltage behind the transient reactance vector (internally generated voltage) whose magnitude depends on the magnitude of the rotor flux vector, $\boldsymbol{\psi}_r$. Although $\boldsymbol{\psi}_r$ is dependent on the generator stator and rotor currents, it can also be manipulated by adjustment of the rotor voltage vector, $\boldsymbol{V}_r$.

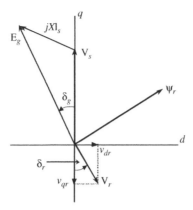

Figure 5.7   Vector diagram of the DFIG operating conditions

Combining Eqs (4.26) and (4.27), the equation for the internal voltage, E_g, may be expressed in vector form as

$$\frac{d}{dt}\overline{E}_g = -\frac{\omega_b}{\overline{T}_0}[\overline{E}_g - j(\overline{X} - \overline{X}')\overline{I}_s] + js\omega_s\overline{E}_g - j\omega_s\frac{\overline{L}_m}{\overline{L}_{rr}}\overline{V}_r \qquad (5.12)$$

where $\overline{E}_g = \bar{e}_d + j\bar{e}_q$, $\overline{I}_s = \bar{i}_{ds} + j\bar{i}_{qs}$ and $\overline{V}_r = \bar{v}_{dr} + j\bar{v}_{qr}$. In the steady state, $d\overline{E}_g/dt = 0$, so that

$$0 = -\frac{1}{\overline{\omega}_s\overline{T}_0}[\overline{E}_g - j(\overline{X} - \overline{X}')\overline{I}_s] + js\overline{E}_g - j\frac{\overline{L}_m}{\overline{L}_{rr}}\overline{V}_r \qquad (5.13)$$

In Eq. (5.13), for normal operating values of s (where the DFIG rotor speed is distinct from the synchronous value), the term having the divider $\overline{\omega}_s\overline{T}_0$ is small compared with the final two terms, so that Eq. (5.13) is reduced to the approximate relationship

$$s\overline{E}_g \approx \frac{\overline{L}_m}{\overline{L}_{rr}}\overline{V}_r; \quad \frac{\overline{L}_m}{\overline{L}_{rr}} = 0.975 \quad \text{(2 MW DFIG)} \qquad (5.14)$$

In Eq. (5.14), as the term $\overline{L}_m/\overline{L}_{rr}$ has a value close to unity (for a 2 MW DFIG wind turbine – see data in Appendix D; for other machine models, this value may differ), the rotor voltage vector V_r is given approximately as $\overline{V}_r \approx s\overline{E}_g$. Hence, since the magnitude of the internal voltage, $|E_g|$, varies only slightly, the magnitude of the rotor voltage is approximately proportional to the slip magnitude. Further, for sub-synchronous operation where slip, s, is positive, V_r is approximately in-phase with the internal voltage vector E_g and for super-synchronous operation, where s is negative, the two voltage vectors are approximately in anti-phase (Figure 5.7).

The angle δ_g in Figure 5.7, which defines the position of the internally generated voltage vector, $\boldsymbol{E}_g$, with respect to the stator voltage, $\boldsymbol{V}_s$, (and hence the q axis of the reference frame), is determined by the power output of the generator. Since the internally generated voltage vector, $\boldsymbol{E}_g$, is orthogonal to the rotor flux vector, $\boldsymbol{\psi}_r$, the angle between the rotor flux vector, $\boldsymbol{\psi}_r$, and the d axis of the reference frame is also given by δ_g.

5.3   Control for Optimum Wind Power Extraction

Dynamic control of the DFIG is provided through the power converter, which permits variable-speed operation of the wind turbine by decoupling the power system electrical frequency and the rotor mechanical speed (Ekanayake *et al.*, 2003). One control scheme, implemented by a number of manufacturers and modelled in this chapter, uses the rotor-side converter to provide torque control together with terminal voltage or power factor (PF) control for the overall system, while the network-side converter controls the DC link voltage. In some applications, the network-side converter is used to provide reactive power. In the DFIG control strategies presented in this chapter, the network-side converter is used to maintain the DC bus voltage and to provide a path for rotor power to and from the AC system at unity power factor (Peña *et al.*, 1996; Holdsworth *et al.*, 2003).

The aim of the control strategy is to extract maximum power from the wind. A typical wind turbine characteristic with the optimal power extraction–speed curve plotted to intersect the $C_{p\,max}$ points for each wind speed is shown in Figure 5.8a. The curve P_{opt} defines the maximum energy capture and the control objective is to keep the turbine on this curve as the wind speed varies.

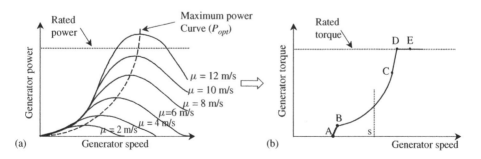

Figure 5.8   Wind turbine characteristic for maximum power extraction

The curve P_{opt} is defined by (Peña *et al.*, 1996; Holdsworth *et al.*, 2003)

$$P_{opt} = K_{opt}\omega_r^3 \tag{5.15}$$

or

$$T_{opt} = K_{opt}\omega_r^2 \tag{5.16}$$

where T_{opt} is the optimal torque of the machine and K_{opt} is a constant obtained from the aerodynamic performance of the wind turbine (usually provided by the manufacturer). The complete generator torque–speed characteristic, which is applied for the controller model is shown in Figure 5.8b. For optimal power extraction, the torque–speed curve is characterized by Eq. (5.16). This is between points B and C. Within this operating range, during low to medium wind speeds, the maximum possible energy can then be extracted from the turbine.

Due to power converter ratings, it is not practical to maintain optimum power extraction over all wind speeds. Therefore, for very low wind speeds the model operates at almost constant rotational speed (A–B). The rotational speed is also often limited by aerodynamic noise constraints, at which point the controller allows the torque to increase, at essentially constant speed (C–D) until rated torque. If the wind speed increases further to exceed the turbine torque rating, the control objective follows D–E, where the electromagnetic torque is constant. When the system reaches point E, pitch regulation takes over from the torque control to limit aerodynamic input power. For very high wind speeds, the pitch control will regulate the input power until the wind speed shutdown limit is reached.

5.4   Control Strategies for a DFIG

5.4.1   Current-mode Control (PVdq)

This technique is often used for the electrical control of the DFIG (Peña *et al.*, 1996). The rotor current is split into two orthogonal components, d and q. The q component of the current is used to regulate the torque and the d component is used to regulate power factor or terminal voltage. For convenience, this controller is termed PVdq control in this book.

5.4.1.1   Torque Control Scheme

The purpose of the torque controller is to modify the electromagnetic torque of the generator according to wind speed variations and drive the system to the required operating point reference. Given a rotor speed measurement, the reference torque provided by the wind turbine characteristic for maximum power extraction (Figure 5.8b) is manipulated to generate a reference value for the rotor current in the q axis, $i_{qr_{ref}}$. The rotor voltage v_{qr} required to operate at the reference torque set point is obtained through a PI controller and the summation of a compensation term to minimize cross-coupling between speed and voltage control loops.

Using the expressions for the voltage behind a transient reactance, $\bar{e}_d$ and $\bar{e}_q$ [Eqs (4.19) and (4.20)] and rotor flux equations [Eqs (4.13) and (4.14)], the torque [given in Eq. (4.38)] can be expressed as

$$\overline{T}_e = \overline{L}_m(\bar{i}_{dr}\bar{i}_{qs} - \bar{i}_{qr}\bar{i}_{ds}) \tag{5.17}$$

From Eq. (4.7), neglecting the stator resistance and stator transients and substituting for ψ_{qs} from Eq. (4.12), then

$$\bar{v}_{ds} = -\overline{\omega}_s(-\overline{L}_{ss}\bar{i}_{qs} + \overline{L}_m\bar{i}_{qr}) \tag{5.18}$$

From this equation, an expression is obtained for $\bar{i}_{qs}$:

$$\bar{i}_{qs} = \frac{1}{\overline{\omega}_s\overline{L}_{ss}}\bar{v}_{ds} + \frac{\overline{L}_m}{\overline{L}_{ss}}\bar{i}_{qr} \tag{5.19}$$

Due to the stator flux oriented (SFO) reference frame used, the d-axis component of the stator voltage $\bar{v}_{ds} = 0$ and Eq. (5.19) can be reduced to

$$\bar{i}_{qs} = \frac{\overline{L}_m}{\overline{L}_{ss}}\bar{i}_{qr} \tag{5.20}$$

Now, from Eq. (4.8), neglecting the stator resistance and stator transients and then substituting ψ_{ds} from Eq. (4.11), we obtain

$$\bar{v}_{qs} = \overline{\omega}_s(-\overline{L}_{ss}\bar{i}_{ds} + \overline{L}_m\bar{i}_{dr}) \tag{5.21}$$

From Eq. (5.21), $\bar{i}_{ds}$ can be expressed as

$$\bar{i}_{ds} = -\frac{1}{\overline{\omega}_s\overline{L}_{ss}}\bar{v}_{qs} + \frac{\overline{L}_m}{\overline{L}_{ss}}\bar{i}_{dr} \tag{5.22}$$

Now, substitution of Eqs (5.20) and (5.22) in Eq. (5.17) gives

$$\overline{T}_e = \overline{L}_m\left[\bar{i}_{dr}\left(\frac{\overline{L}_m}{\overline{L}_{ss}}\bar{i}_{qr}\right) - \bar{i}_{qr}\left(-\frac{1}{\overline{\omega}_s\overline{L}_{ss}}\bar{v}_{qs} + \frac{\overline{L}_m}{\overline{L}_{ss}}\bar{i}_{dr}\right)\right] \tag{5.23}$$

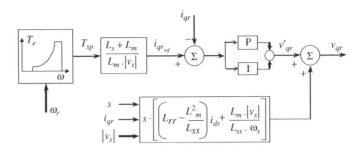

Figure 5.9   DFIG torque control strategy

After simplifying Eq. (5.23), the reference for $\bar{i}_{qr}$ to obtain the desired optimal torque is expressed as

$$\bar{i}_{qr_{\text{ref}}} = \frac{\bar{\omega}_s \bar{L}_{ss}}{\bar{L}_m \bar{v}_{qs}} \bar{T}_{sp} \tag{5.24}$$

where $\bar{T}_{sp}$ stands for the optimal torque set point provided from the torque–speed characteristic for maximum power extraction. A block diagram of this control scheme is shown in Figure 5.9. Although the over-bar notation to identify per unit quantities has been omitted, all the variables shown in the block diagram are in per unit.

The difference in the rotor current, $\bar{i}_{qr}$, from the reference value, $\bar{i}_{qr_{\text{ref}}}$, forms the error signal that is processed by the PI compensator to produce the rotor voltage, $\bar{v}'_{qr}$. To obtain the required value of the rotor voltage in the q-axis, $\bar{v}_{qr}$, a compensation term is added to the PI compensator to minimize the cross-coupling between torque and voltage control loops.

From Eq. (4.10), neglecting the transient term and substituting for ψ_{dr} from Eq. (4.13), then

$$\bar{v}_{qr} = \bar{r}_r \bar{i}_{qr} + s\bar{\omega}_s (\bar{L}_{rr} \bar{i}_{dr} - \bar{L}_m \bar{i}_{ds}) \tag{5.25}$$

By substituting for $\bar{i}_{ds}$ from Eq. (5.22) into Eq. (5.25), the following equation can be obtained:

$$\bar{v}_{qr} = \bar{r}_r \bar{i}_{qr} + s\bar{\omega}_s \left(\bar{L}_{rr} - \frac{\bar{L}_m^2}{\bar{L}_{ss}} \right) \bar{i}_{dr} - \frac{\bar{L}_m}{\bar{\omega}_s \bar{L}_{ss}} \bar{v}_{qs} \tag{5.26}$$

The expression for the compensation term is given by the second and third terms on the right-hand of Eq. (5.26).

5.4.1.2 Voltage Control Scheme

The strategy for voltage control is typically designed to provide terminal voltage or power factor control using the rotor-side converter. Although reactive power injection can also be obtained from the network-side converter, for DFIG voltage control schemes the rotor-side converter is likely to be preferred to the network-side converter. The reactive power through the rotor, Q_r, is given as

$$Q_r = \text{Im}[\mathbf{V}_r \mathbf{I}_r^*] \tag{5.27}$$

and then, when Q_r is referred to the stator:

$$Q_r' = \text{Im}\left[\frac{\mathbf{V}_r \mathbf{I}_r^*}{s}\right] \tag{5.28}$$

As shown by Eq. (5.28), the reactive power injection through the rotor circuit is effectively amplified by a factor of $1/s$. This is the main reason why the rotor-side converter is the preferred option to provide the machine requirements for reactive power.

The control action for terminal voltage or power factor control is derived as follows (Holdsworth *et al.*, 2003). Consider the total grid (stator) side reactive power in per unit given by

$$\overline{Q}_s = \overline{Q}_{\text{grid}} = \text{Im}(\overline{v}_s \overline{i}_s^*) = \overline{v}_{qs}\overline{i}_{ds} - \overline{v}_{ds}\overline{i}_{qs} \tag{5.29}$$

As a result of the SFO reference frame used to develop the mathematical model of the DFIG, the d-axis component of the stator voltage $\overline{v}_{ds} = 0$. Using the stator flux equations, the stator current in the d-axis, $\overline{i}_{ds}$, can be expressed as

$$\overline{i}_{ds} = -\frac{1}{\overline{L}_{ss}}\overline{\psi}_{ds} + \frac{\overline{L}_m}{\overline{L}_{ss}}\overline{i}_{dr} \tag{5.30}$$

Substituting Eq. (5.30) in Eq. (5.29):

$$\overline{Q}_{\text{grid}} = \overline{v}_{qs}\left(-\frac{1}{\overline{L}_{ss}}\overline{\psi}_{ds} + \frac{\overline{L}_m}{\overline{L}_{ss}}\overline{i}_{dr}\right) \tag{5.31}$$

From the stator voltage equation and neglecting the stator resistance, the flux linkage $\overline{\psi}_{ds}$ can be expressed as

$$\overline{\psi}_{ds} = \frac{\overline{v}_{qs}}{\overline{\omega}_s} \tag{5.32}$$

Substituting Eq. (5.32) in Eq. (5.31) gives

$$\overline{Q}_{grid} = -\frac{\overline{v}_{qs}^{\,2}}{\overline{\omega}_s \overline{L}_{ss}} + \frac{\overline{L}_m \overline{v}_{qs}}{\overline{L}_{ss}} \overline{i}_{dr} \tag{5.33}$$

The rotor current component $\overline{i}_{dr}$ is divided into a generator magnetizing component $\overline{i}_{dr\_m}$ and a component for controlling reactive power flow (or terminal voltage) with the connecting network $\overline{i}_{dr\_g}$. The total reactive power is also divided into $\overline{Q}_{mag}$ and $\overline{Q}_{gen}$. Hence Eq. (5.33) can be now expressed as

$$\overline{Q}_{grid} = \overline{Q}_{mag} + \overline{Q}_{gen} = -\frac{\overline{v}_{qs}^{\,2}}{\overline{\omega}_s \overline{L}_{ss}} + \frac{\overline{L}_m \overline{v}_{qs}}{\overline{L}_{ss}} (\overline{i}_{dr\_m} + \overline{i}_{dr\_g}) \tag{5.34}$$

from which

$$\overline{Q}_{mag} = -\frac{\overline{v}_{qs}^{\,2}}{\overline{\omega}_s \overline{L}_{ss}} + \frac{\overline{L}_m \overline{v}_{qs}}{\overline{L}_{ss}} \overline{i}_{dr\_m} \tag{5.35}$$

To compensate for the no-load reactive power absorbed by the machine, $\overline{Q}_{mag}$ must equate to zero. To obtain this, from Eq. (5.35) the value of $\overline{i}_{dr\_m}$ is controlled to equal

$$\overline{i}_{dr\_m} = \frac{\overline{v}_{qs}}{\overline{\omega}_s \overline{L}_m} \tag{5.36}$$

As the terminal voltage will increase or decrease when more or less reactive power is delivered to the grid, the voltage control should fulfil the following requirements: (i) the reactive power consumed by the DFIG should be compensated by i_{dr_m} and (ii) if the terminal voltage is too low or too high compared with the reference value then i_{dr_g} should be adjusted appropriately.

A block diagram of the DFIG terminal voltage controller is shown in Figure 5.10. The required rotor voltage in the d-axis, $\overline{v}_{dr}$, is obtained through

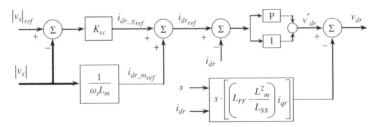

Figure 5.10 DFIG terminal voltage control strategy (the control gain K_{vc} is adjusted to improve terminal voltage or power factor performance)

the output of a PI controller, $\bar{v}'_{dr}$, minus a compensation term to eliminate cross-coupling between control loops. In this case, the compensation term is derived from the equation of the rotor voltage in the d-axis. All variables shown in Figure 5.10 are in per unit.

The operation of the rotor-side converter with regard to the terminal voltage or power factor control is entirely dependent upon the requirements or the preferred operation of the system. If the rotor current is required to be kept to a minimum, such that $\bar{i}_{dr_{ref}} = 0$, the current drawn by the machine to maintain the field flux will be provided by the d-axis component of the stator current, $\bar{i}_{ds}$. Therefore, the voltage at the terminals will be reduced resulting from the reactive power absorbed by the machine. To implement this strategy, the rotor current reference, shown in Figure 5.10, should be set to zero, i.e. $\bar{i}_{dr_{ref}} = 0$. If the rotor voltage, $\bar{v}_{dr}$, obtained from the rotor-side converter is used to control the DFIG terminal voltage/power factor and to maintain a constant machine field flux, the magnitude of the rotor current, $\bar{i}_{dr}$, will not be zero. Then, as shown previously, the d-axis component of the rotor current reference value is split into a part that magnetizes the generator, $\bar{i}_{dr\_m}$, which effectively controls the power factor of the machine, and a part that determines the net reactive power exchange with the grid, $\bar{i}_{dr\_g}$.

5.4.2 Rotor Flux Magnitude and Angle Control

This control methodology exercises control over the generator terminal voltage and power output by adjusting the magnitude and angle of the rotor flux vector (Hughes *et al.*, 2005; Anaya-Lara *et al.*, 2006). This strategy has the advantage of providing low interaction between the power and voltage control loop and enhanced system damping and voltage recovery following faults. The structure of the flux magnitude and angle controller (FMAC) is illustrated in Figure 5.11. It comprises two distinct loops, one to control the terminal voltage and the other to control the power output of the generator. Since the DFIG internal voltage vector, $\boldsymbol{E}_g$, is directly related to the rotor flux vector, $\boldsymbol{\psi}_r$, either of these vectors can be employed as control vector. In the following discussion and examples, $\boldsymbol{E}_g$ has been selected as control vector.

5.4.2.1 Voltage Control

In the voltage control loop, the difference in the magnitude of the terminal voltage, V_s, from its desired reference value, $V_{s_{ref}}$, forms an error signal that is processed via the AVR compensator to produce the reference value for the magnitude of the DFIG internal voltage vector, $|\boldsymbol{E}_g|_{ref}$.

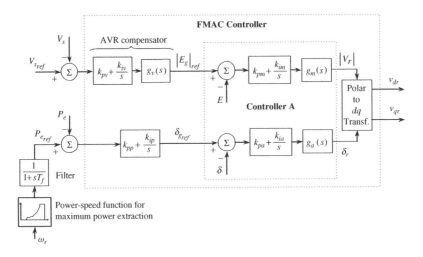

Figure 5.11   Block diagram of the FMAC controller

5.4.2.2   Power Control

In the power control loop, the reference set point value, $P_{e_{ref}}$, is determined by the wind turbine power–speed characteristic for maximum power extraction from the prevailing wind velocity (Müller *et al.*, 2002). The difference in the generator power, P_e, from the reference set point value, $P_{e_{ref}}$, forms the basic error signal that is processed by the compensator to produce the reference value for the angular position of the control vector, $\delta_{g_{ref}}$, with respect to the stator voltage vector.

Both the voltage and power control loops employ PI controllers, with the provision of additional lead–lag compensation in the case of the voltage loop to ensure suitable margins of loop stability.

Controller A employs the reference signals, $|E_g|_{ref}$ and $\delta_{g_{ref}}$, to provide the magnitude and angle of the rotor voltage vector, V_r. PI control with additional lead–lag compensation is employed to provide appropriate speed of response and stability margins in the individual loops.

The rotor voltage vector, V_r, is then transformed from its polar coordinates to rectangular dq coordinates v_{dr} and v_{qr} and used by the PWM generators to control the switching operation of the rotor-side converter. Typical control parameters and transfer functions can be found in Appendix D.

5.5   Dynamic Performance Assessment

The operation and dynamic characteristics of the DFIG with current-mode control (PVdq) are assessed using the simple circuit shown in Figure 5.12,

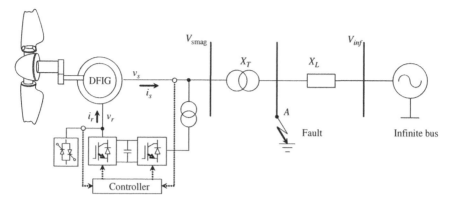

Figure 5.12 Network model used to assess the performance of a DFIG with the current-mode controller

where the DFIG is connected to an infinite bus through the impedances of the turbine transformer, X_T, and the transmission line, X_L. A four-pole 2 MW DFIG represented by a third-order model is used in this case.

5.5.1 Small Disturbances

In the following examples, small step changes in the mechanical input torque and small variations in the torque and voltage set points of the DFIG controller are considered.

5.5.1.1 Step Change in Mechanical Torque Input

This example illustrates an operating condition where the available aerodynamic power reduces due to a decrease in the wind speed. For this test, the mechanical input torque applied to the wind turbine is decreased by 20% at $t = 1$ s (Figure 5.13). The DFIG operates initially in super synchronous mode with a slip of $s = -0.2$ pu. The mechanical torque corresponding to this rotor speed is $T_m = 0.8064$ pu. (obtained from the turbine characteristic for maximum power extraction). The DFIG responses for this scenario are shown in Figure 5.14.

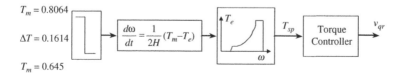

Figure 5.13 Step change in the mechanical torque input

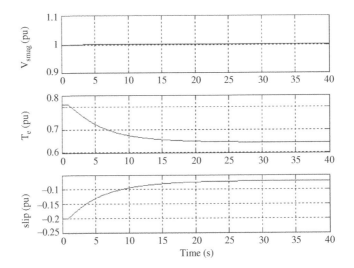

Figure 5.14 DFIG responses for a 20% decrease in the mechanical torque input at $t = 1$ s (initial torque $T_m = T_e = 0.8064$ pu). New slip, $s = -0.0706$ pu. New torque, $T_m = T_e = 0.645$ pu

In Figure 5.14, V_{smag} is the terminal voltage of the DFIG and T_e is the torque output. When the mechanical torque is reduced at $t = 1$ s, the torque output of the generator also reduces until a new operating condition is reached at approximately $t = 30$ s. In this new operating point, the slip equals $s = -0.0706$ pu and the electrical torque settles at $T_e = 0.645$ pu to match the mechanical input torque. In this case, the large lumped turbine, shaft and generator rotor inertia dominates the dynamic control performance of the DFIG.

Although the torque of the generator is significantly modified, the impact that this change has on the generator terminal voltage, V_{smag}, is minimal.

5.5.1.2 Step Change in the Torque Reference Value

This example illustrates the case where the operating point of the generator is adjusted in the event of small disturbances in the power network such as light load variations. The mechanical torque input applied to the wind turbine is kept constant and a small step change is applied at $t = 1$ s in the torque reference, T_{sp}, of the torque control loop (Figure 5.15).

At $t = 1$ s, the torque set point is increased by 20% (which corresponds to an increase of 0.16 pu). The DFIG is initially operating in super synchronous mode with a slip of $s = -0.2$ pu and $T_e = 0.8064$. The responses for this operation are shown in Figure 5.16. When the electrical torque reference is

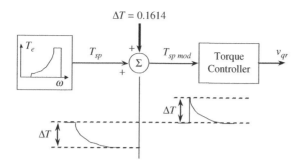

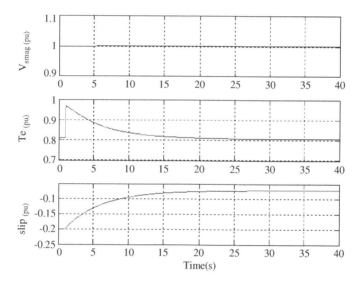

Figure 5.15 Step change in the mechanical torque input

Figure 5.16 DFIG responses for a 20% increase (disturbance) in the torque reference value T_{sp}

increased at $t = 1$ s, a mismatch between mechanical and electrical torque is developed and therefore the DFIG slows. Initially, the electrical torque output increases sharply due to the action of the torque controller, which tries to follow the new torque reference value. However, as the mechanical torque input (aerodynamic power) is held fixed, the torque controller adjusts the speed of the generator to provide a torque set point, T_{sp}, which added to the 20% increase provides a new torque reference, $T_{sp\,mod}$, that matches the mechanical torque input of $T_e = 0.8064$ pu as shown in Figure 5.17. The slip settles at $s = -0.706$ pu, which agrees with the previous example (Figure 5.14).

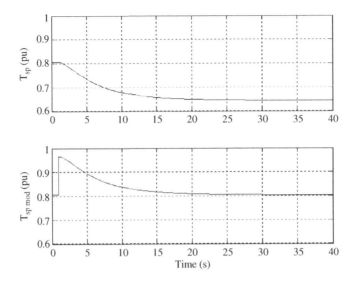

Figure 5.17 Torque set point input to the DFIG controller. A 20% increase (disturbance) is applied in the torque reference value at $t = 1$ s

5.5.1.3   Step Change in the Voltage Reference Value

In this example, a step increase is applied in the voltage reference value of the voltage control loop at $t = 1$ s and removed after 4 s. The DFIG responses are shown in Figure 5.18, where the correct performance of the voltage controller is observed.

5.5.2   Performance During Network Faults

In the following example, a balanced three-phase fault (short-circuit) is applied at the high-voltage terminals of the DFIG transformer (point A in Figure 5.12). The fault is applied at $t = 1$ s and cleared after 150 ms. The results are shown without crowbar protection. The converter is assumed to be sufficiently robust to provide all the demands of the DFIG controller during transient operation (i.e. the converters withstand the fault current developed in the experiment).

The responses of the DFIG terminal voltage, V_{smag}, electrical torque, T_e, and slip, s, are shown in Figure 5.19. Due to the fault, the electrical torque of the generator falls to zero and therefore the machine starts to speed up. During the fault, the terminal voltage drops to approximately 0.45 pu (retained voltage). When the fault is cleared after 150 ms, the system recovers stability and the voltage recovers the pre-fault state with a fast and smooth response.

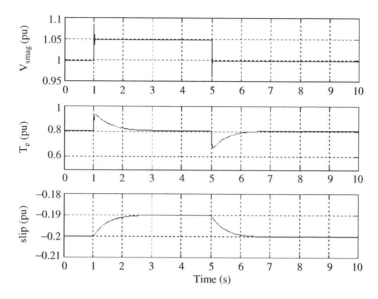

Figure 5.18 DFIG responses for a 5% increase (disturbance) in the voltage reference value
(V_{ref}) at $t = 1$ s. At $t = 5$ s, the step disturbance is removed

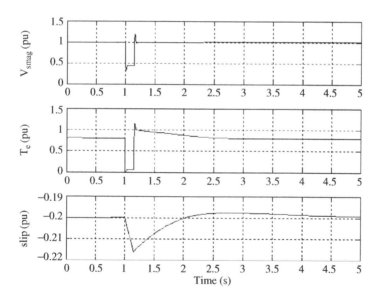

Figure 5.19 DFIG responses for a fault applied at $t = 1$ s with a duration of 150 ms; slip
$s = -0.216$. Crowbar protection is not in operation

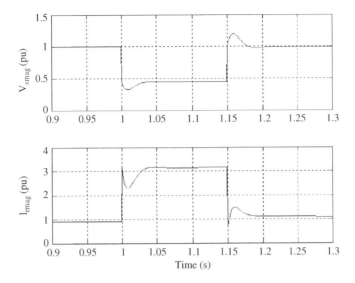

Figure 5.20 DFIG responses for a fault applied at $t = 1$ s with a fault clearance time of 150 ms. $V_{smag} = 0.45$ pu; $I_{rmag} = 3.15$ pu. Crowbar protection is not in operation

Figure 5.20 shows a snapshot (0.9–1.3 s) of the terminal voltage and rotor current, I_{rmag}, at the instant of the fault. As the crowbar protection is not in operation, the rotor current increases freely up to a value of approximately $I_{rmag} = 3.15$ pu. Without crowbar protection, the maximum value that the rotor current can reach during the fault is then just limited by the DFIG and power network parameters.

References

Anaya-Lara, O., Hughes, F. M., Jenkins, N. and Strbac, G. (2006) Rotor flux magnitude and angle control strategy for doubly fed induction generators, *Wind Energy*, **9** (5), 479–495.

Ekanayake, J. B., Holdsworth, L., Wu, X. and Jenkins, N. (2003) Dynamic modelling of doubly fed induction generator wind turbines, *IEEE Transactions on Power Systems*, **18** (2), 803–809.

Fox, B., Flynn, D., Bryans, L., Jenkins, N., Milborrow, D., OMalley, M., Watson, R. and Anaya-Lara, O. (2007) *Wind Power Integration: Connect and System Operational Aspects"*, IET Power and Energy Series, Vol. 50, Institution of Engineering and Technology, Stevenage, ISBN 10: 0863414494.

Hindmarsh, J. (1995) Electrical Machines and Their Applications, 4th edn, Butterworth-Heinemann, Oxford.

Holdsworth, L., Wu, X., Ekanayake, J. B. and Jenkins, N. (2003) Comparison of fixed speed and doubly-fed induction wind turbines during power system disturbances, *IEE Proceedings Generation, Transmission and Distribution*, **150** (3), 343–352.

Hughes, F. M., Anaya-Lara, O., Jenkins, N. and Strbac, G. (2005) Control of DFIG-based wind generation for power network support, *IEE Transactions on Power Systems*, **20** (4), 1958–1966.

Müller, S., Deicke, M. and De Doncker, R. W. (2002) Doubly fed induction generator systems for wind turbines, *IEE Industry Applications Magazine*, **8** (3), 26–33.

Peña, R., Clare, J. C. and Asher, G. M. (1996) Doubly fed induction generator using back-to-back PWM converters and its application to variable speed wind-energy generation, *IEE Proceedings Electrical Power Applications*, **143** (3), 231–241.

6

Fully Rated Converter-based (FRC) Wind Turbines

In order to fulfil present Grid Code requirements, wind turbine manufacturers have been considering induction or synchronous generators with fully rated voltage source converters to give full-power, converter-controlled, variable-speed operation. This chapter describes the main components and features of these technologies and presents results from recent studies conducted on their dynamic modelling and control design.

The typical configuration of a fully rated converter-based (FRC) wind turbine is shown in Figure 1.8. This type of wind turbine may or may not have a gearbox and a wide range of electrical generator types such as asynchronous, conventional synchronous and permanent magnet can be employed. As all the power from the wind turbine is transferred through the power converter, the specific characteristics and dynamics of the electrical generator are effectively isolated from the power network (Fox *et al.*, 2007). Hence the electrical frequency of the generator may vary as the wind speed changes, while the network frequency remains unchanged, permitting variable-speed operation. The rating of the power converter in this wind turbine corresponds to the rated power of the generator.

The power converter can be arranged in various ways. While the generator-side converter (GSC) can be a diode-based rectifier or a PWM voltage source converter, the network-side converter (NSC) is typically a PWM voltage source converter. The strategy to control the operation of the generator and power flows to the network depend very much on the type of power converter arrangement employed.

Wind Energy Generation: Modelling and Control   Olimpo Anaya-Lara, Nick Jenkins,
Janaka Ekanayake, Phill Cartwright and Mike Hughes
© 2009 John Wiley & Sons, Ltd

6.1 FRC Synchronous Generator-based (FRC-SG) Wind Turbine

In an FRC wind turbine based on synchronous generators, the generator can be electrically excited or it can have a permanent magnet rotor. In the direct-drive arrangement, the turbine and generator rotors are mounted on the same shaft without a gearbox and the generator is specially designed for low-speed operation with a large number of poles. The synchronous generators of direct-drive turbines tend to be very large due to the large number of poles. However, if the turbine includes a gearbox (typically a single-stage gearbox with low ratio), then a smaller generator with a smaller number of poles can be employed (Akhmatov *et al.*, 2003).

6.1.1 Direct-driven Wind Turbine Generators

Today, almost all wind turbines rated at a few kilowatts or more use standard (four pole) generators for speeds between 750 and 1800 rpm. The turbine speed is much lower than the generator speed, typically between 20 and 60 rpm. Therefore, in a conventional wind turbine, a gearbox is used between the turbine and the generator. An alternative is to use a generator for very low speeds. The generator can then be directly connected to the turbine shaft. The drive trains of a conventional wind turbine and one with a direct-driven generator are shown schematically in Figure 6.1 (Grauers, 1996).

There are two main reasons for using direct-driven generators in wind turbine systems. Direct-driven generators are favoured for some applications due to reduction in losses in the drive train and less noise (Grauers, 1996).

The most important difference between conventional and direct-driven wind turbine generators is that the low speed of the direct-driven generator makes a very high-rated torque necessary. This is an important difference, since the size

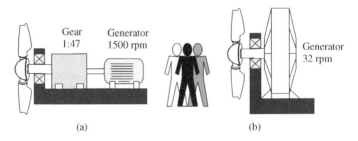

(a)   (b)

Figure 6.1 Drive trains of (a) a conventional wind turbine and (b) one with a direct-drive generator (Grauers, 1996)

and the losses of a low-speed generator depend on the rated torque rather than on the rated power. A direct-driven generator for a 500 kW, 30 rpm wind turbine has the same rated torque as a 50 MW, 3000 rpm steam-turbine generator.

Because of the high-rated torque, direct-driven generators are usually heavier and less efficient than conventional generators. To increase the efficiency and reduce the weight of the active parts, direct-driven generators are usually designed with a large diameter. To decrease the weight of the rotor and stator yokes and to keep the end-winding losses small, direct-driven generators are also usually designed with a small pole pitch.

6.1.2 Permanent Magnets Versus Electrically Excited Synchronous Generators

The synchronous machine has the ability to provide its own excitation on the rotor. Such excitation may be obtained by means of either a current-carrying winding or permanent magnets (PMs). The wound-rotor synchronous machine has a very desirable feature compared with its PM counterpart, namely an adjustable excitation current and, consequently, control of its output voltage independent of load current. This feature explains why most constant-speed, grid-connected hydro and turbo generators use wound rotors instead of PM-excited rotors. The synchronous generator in wind turbines is in most cases connected to the network via an electronic converter. Therefore, the advantage of controllable no-load voltage is not as critical.

Wound rotors are heavier than PM rotors and typically bulkier (particularly in short pole-pitch synchronous generators). Also, electrically excited synchronous generators have higher losses in the rotor windings. Although there will be some losses in the magnets caused by the circulation of eddy currents in the PM volume, they will usually be much lower than the copper losses of electrically-excited rotors. This increase in copper losses will also increase on increasing the number of poles.

6.1.3 Permanent Magnet Synchronous Generator

PM excitation avoids the field current supply or reactive power compensation facilities needed by wound-rotor synchronous generators and induction generators and it also removes the need for slip rings (Chen and Spooner, 1998). Figure 6.2 shows the arrangement with an uncontrolled diode-based rectifier as the generator-side converter. A DC booster is used to stabilize the DC link voltage whereas the network-side converter (PWM-VSC) controls the operation of the generator. The PWM-VSC can be controlled using load-angle techniques or current controllers developed in a voltage-oriented *dq* reference

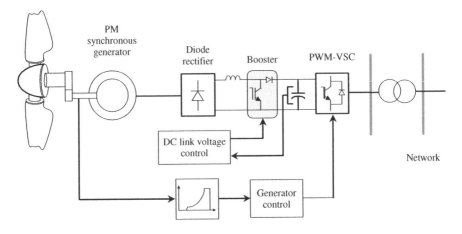

Figure 6.2 Permanent magnet synchronous generator with diode rectifier

frame. The power reference is defined by the maximum power–speed characteristic shown in Figure 5.8 with speed and power limits.

The topology with a permanent magnet synchronous generator and a power converter system consisting of two back-to-back voltage source converters is illustrated in Figure 6.3. In this arrangement, the generator-side converter controls the operation of the generator and the network-side converter controls the DC link voltage by exporting active power to the network.

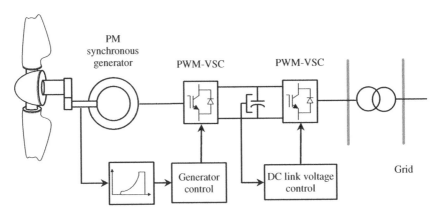

Figure 6.3 Permanent magnet synchronous generator with two back-to-back voltage source converters

6.1.4 Wind Turbine Control and Dynamic Performance Assessment

Control over the power converter system can be exercised with different schemes. The generator-side converter can be controlled using load angle control techniques or using vector control. The network-side converter is commonly controlled using load angle control techniques.

6.1.4.1 Generator-side Converter Control and Dynamic Performance

The generator-side converter controls the operation of the wind turbine and two control techniques are explained, namely load angle and vector control.

Load Angle Control Technique
The load angle control strategy employs steady-state power flow equations (Kundur, 1994; Fox *et al.*, 2007) to determine the transfer of active and reactive power between the generator and the DC link. With reference to Figure 6.4, E_g is the magnitude of the generator internal voltage, X_g the synchronous reactance, V_t the voltage (magnitude) at the converter terminals and α_g is the phase difference between the voltages E_g and V_t.

The active and reactive power flows in the steady state are defined as

$$P = \frac{E_g V_t}{X_g} \sin \alpha_g \tag{6.1}$$

$$Q = \frac{E_g{}^2 - E_g V_t \cos \alpha_g}{X_g} \tag{6.2}$$

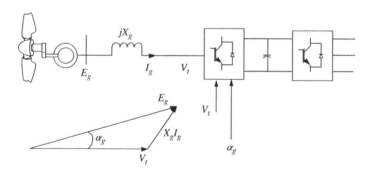

Figure 6.4 Load angle control of a synchronous generator wind turbine

As the load angle α_g is generally small, $\sin\alpha_g \approx \alpha_g$ and $\cos\alpha_g \approx 1$. Hence Eqs (6.1) and (6.2) can be simplified to

$$P = \frac{E_g V_t}{X_g}\alpha_g \tag{6.3}$$

$$Q = \frac{E_g^2 - E_g V_t}{X_g} \tag{6.4}$$

From Eqs (6.3) and (6.4), it can be seen that the active power transfer depends mainly on the phase angle α_g. The reactive power transfer depends mainly on voltage magnitudes and it is transmitted from the point with higher voltage magnitude to the point with lower magnitude.

The operation of the generator and the power transferred from the generator to the DC link are controlled by adjusting the magnitude and angle of the voltage at the AC terminals of the generator-side converter. The magnitude, V_t, and angle, α_g, required at the terminal of the generator-side converter are calculated using Eqs (6.3) and (6.4) as

$$\alpha_g = \frac{P_{gref} X_g}{E_g V_t} \tag{6.5}$$

$$V_t = E_g - \frac{Q_{gref} X_g}{E_g} \tag{6.6}$$

where P_{gref} is the reference value of the active power that needs to be transferred from the generator to the DC link and Q_{gref} is the reference value for the reactive power.

The reference value P_{gref} is obtained from the maximum power extraction curve (Figure 5.8) for a given generator speed, ω_r. As the generator has permanent magnets, it does not require a magnetizing current through the stator, hence the reactive power reference value can be set to zero, $Q_{gref} = 0$ (i.e. V_t and E_g are equal in magnitude). The implementation of the load angle control scheme is shown in Figure 6.5.

The major advantage of the load angle control is its simplicity. However, as in this technique the dynamics of the generator are not considered, it may not be very effective in controlling the generator during a transient operating condition.

Dynamic Performance Assessment The performance of the load angle control strategy is illustrated using steady-state, reduced order and non-reduced order models of the synchronous generator to explore the influence that generator

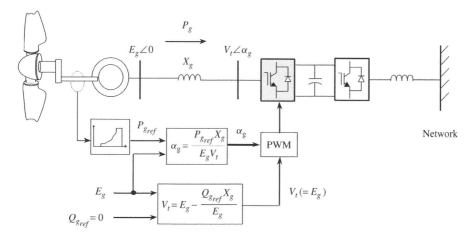

Figure 6.5   Load angle control of the generator-side converter

stator and rotor transients have on control performance. The test system used for the simulations is that shown in Figure 6.5, where the wind turbine is connected to an infinite busbar with parameters given in Appendix D. The mechanical structure of the turbine is represented by a single-mass model and ideal operation of the voltage source converters is assumed.

A step change increase in the mechanical torque input from 60×10^3 to 80×10^3 N m is applied at 10 s and then it is decreased back to the initial value $(60 \times 10^3$ N m) at 20 s. The electromagnetic torques of the non-reduced order, reduced order and steady-state generator models are given in Figure 6.6. It can be seen that the load angle control tracks satisfactorily the torque reference with the three models of the synchronous generator. The responses of the electrical speed and load angle are also given in Figure 6.6.

Since the steady-state model neglects the stator and rotor transients, the responses obtained with this model reach the final value more rapidly (Figure 6.6). Although both reduced and non-reduced order models in general give similar responses, a significant difference can be seen in the stator current, i_{ds} (Figure 6.7), where current oscillations due to the stator transients appear in the response obtained with the non-reduced order model. These current oscillations are damped out by the damper windings in the generator rotor and are only present during the transient period.

The frequency of the current oscillations observed in the non-reduced order model corresponds to the operating electrical frequency of the generator. For this particular example, the frequency of oscillation is approximately 41.2 Hz, which agrees with the generator electrical frequency, namely $2\pi \times 41.2 =$

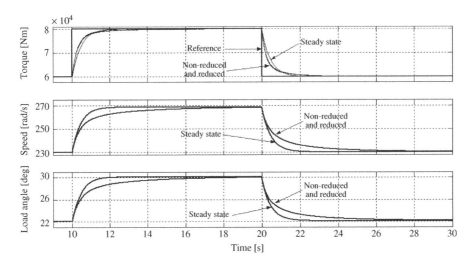

Figure 6.6 Electromagnetic torque, rotor speed (electrical) and load angle variations of non-reduced order, reduced order and steady-state generator models for step changes in input torque with the load angle control strategy. Since the reduced and non-reduced order responses are the same they cannot be distinguished in the figure

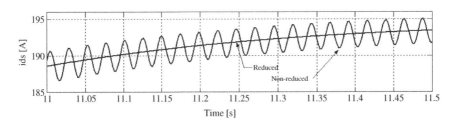

Figure 6.7 Response of the *d* axis stator current with reduced and non-reduced order models of the synchronous generator

258.9 rad s^{-1}, during the time interval shown in Figure 6.7. As the power converter decouples the generator from the network, these oscillations are not transferred to the network. Consequently, the reduced order model may be used as an appropriate representation of the synchronous generator when the load angle control strategy is employed and the overall performance of the variable-speed wind turbine on the power system is the main concern.

Vector Control Strategy

Vector control techniques are implemented based on the dynamic model of the synchronous generator expressed in the *dq* frame. The *dq* frame is defined as the *d* axis aligned with the magnetic axis of the rotor (field).

For the vector control $\bar{i}_{ds_{ref}}$ is set to zero and $\bar{i}_{qs_{ref}}$ is derived from Eq. (3.24). From Eqs (3.24) and (3.27) with $\bar{i}_{ds} = 0$, the following can be obtained:

$$T_e = \overline{\psi}_{ds}\bar{i}_{qs} \tag{6.7}$$

$$\overline{\psi}_{ds} = \overline{L}_{md}\bar{i}_f \tag{6.8}$$

Defining $\overline{L}_{md}\bar{i}_f = \overline{\psi}_{fd}$ and substituting for $\overline{\psi}_{ds}$ from Eq. (6.8) into Eq. (6.7):

$$T_e = \overline{\psi}_{fd}\bar{i}_{qs} \tag{6.9}$$

From Eq. (6.9) for a given torque reference $\overline{T}_{sp}$:

$$\bar{i}_{qs_{ref}} = \frac{\overline{T}_{sp}}{\overline{\psi}_{fd}} \tag{6.10}$$

Once the reference currents, $\bar{i}_{qs_{ref}}$ and $\bar{i}_{ds_{ref}}$, have been determined by the controller, the corresponding voltage magnitudes can be calculated from Eqs (3.31) and (3.33) as

$$\overline{v}_{ds} = -\overline{r}_s\bar{i}_{ds} + \overline{X}_{qs}\bar{i}_{qs} \tag{6.11}$$

$$\overline{v}_{qs} = -\overline{r}_s\bar{i}_{qs} - \overline{X}_{ds}\bar{i}_{ds} + \overline{E}_{fd} \tag{6.12}$$

A PI controller is used to regulate the error between the reference and actual current values, which relate to the $\overline{r}_s$ term in the right-hand side of Eqs (6.11) and (6.12). Additional terms are included to eliminate the cross-coupling effect as shown in Figure 6.8.

The current reference $\bar{i}_{ds_{ref}}$ is kept to zero when the generator operates below the base speed and it is set to a negative value to cancel some of the flux linkage when the generator operates above the base speed. The current reference $\bar{i}_{qs_{ref}}$ is determined from the torque equation. The implementation of the vector control technique is shown in Figure 6.9.

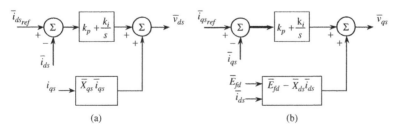

(a) (b)

Figure 6.8 Control loops in the vector control strategy. (a) Magnetizing control loop (d axis); (b) torque control loop (q axis)

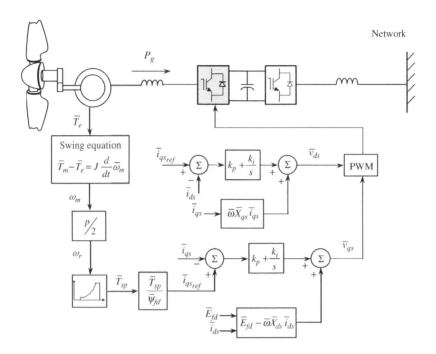

Figure 6.9   Vector control of the generator-side converter

The torque control is exercised in the q axis and the magnetization of the generators is controlled in the d axis. The reference value of the stator current in the q axis, $\bar{i}_{qs_{ref}}$, is calculated from Eq. (6.10) and compared with the actual value, $\bar{i}_{qs}$. The error between these two signals is processed by a PI controller whose output is the voltage in the q axis, $\bar{v}_{qs}$, required to control the generator-side converter. To calculate the required voltage in the d axis, $\bar{v}_{ds}$, the reference value of the stator current in the d axis, $\bar{i}_{ds_{ref}}$, is compared against the actual current in the d axis, $\bar{i}_{ds}$, and the error between these two signals is processed by a PI controller. The reference $\bar{i}_{ds_{ref}}$ may be assumed to be zero for the permanent magnet synchronous generator.

Dynamic Performance Assessment   The performance of the vector control strategy is illustrated using the non-reduced order model of the synchronous generator. A step increase in the torque input from 60×10^3 to 80×10^3 N m is applied at 10 s and at 20 s the torque input is decreased from 80×10^3 back to 60×10^3 N m. The responses of the rotor speed, active power and stator q axis current are shown in Figure 6.10.

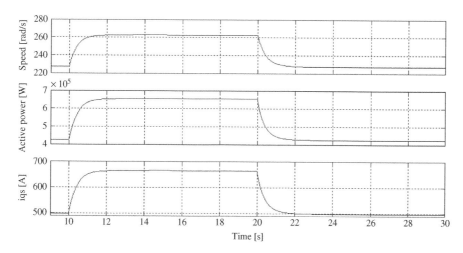

Figure 6.10 Rotor speed (electrical), active power and stator current responses obtained with the vector control strategy for step changes in input torque with non-reduced order generator model

It can be seen that the rotor speed (electrical) and active power achieve the new steady state faster with the vector control strategy than with the load angle control strategy. In the case of the vector control strategy, the transient observed in the responses is associated with the mechanical dynamics of the turbine rather than with electrical transients. The stator current in the q axis increases from 500 to 663 A, which is proportional to the input torque variation.

In the vector control strategy, the controller uses the measured stator currents as feedback signals. For the implementation of this control strategy, the generator model usually includes the stator transients. To demonstrate the importance of the stator transients in the vector control, performance results are shown using the reduced order model of the synchronous generator where the stator transients are neglected. The responses of the torque and rotor speed obtained with both the non-reduced order and reduced order models are shown in Figure 6.11. The results in this figure show that the reduced order model response is oscillatory, in contrast to that obtained with the non-reduced order model. This shows that in order to use the vector control technique it may be necessary to use a non-reduced representation of the synchronous generator to avoid inaccuracies such as those shown during the transient period when the reduced order model was employed.

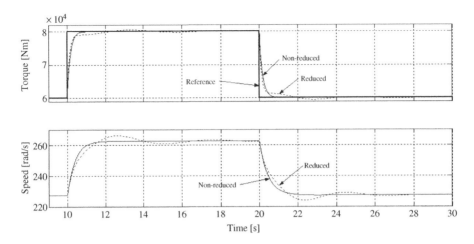

Figure 6.11 Electromagnetic torque and rotor speed (electrical responses of reduced and non-reduced order models of the synchronous generator for step changes in input torque with vector control strategy)

6.1.4.2 Modelling of the DC Link

For simulation purposes, the reference value for the active power, P_{gref}, that needs to be transmitted to the grid can be determined by examining the DC link dynamics with the aid of Figure 6.12. This figure illustrates the power balance at the DC link, which is expressed as

$$P_C = P_g - P_{\text{net}} \tag{6.13}$$

where P_C is the power that goes through the DC link capacitor, C, P_g is the active power output of the generator (and transmitted to the DC link) and P_{net} is the active power transmitted from the DC link to the grid.

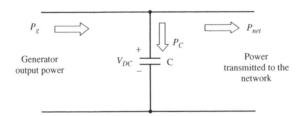

Figure 6.12 Power flow in the DC link

The power flow through the capacitor is given as

$$P_C = V_{DC} I_{DC}$$

$$= V_{DC} C \frac{dV_{DC}}{dt} \tag{6.14}$$

From this equation, the DC link voltage, V_{DC}, is determined as follows:

$$P_C = V_{DC} C \frac{dV_{DC}}{dt} = \frac{C}{2} \times 2 \times V_{DC} \frac{dV_{DC}}{dt}$$

$$= \frac{C}{2} \frac{dV_{DC}^2}{dt} \tag{6.15}$$

Rearranging Eq. (6.15) and integrating both sides of the equation:

$$V_{DC}^2 = \frac{2}{C} \int P_C dt \tag{6.16}$$

then

$$V_{DC} = \sqrt{\frac{2}{C} \int P_C dt} \tag{6.17}$$

By substituting P_C in Eq. (6.17) using Eq. (6.13), the DC link voltage, V_{DC}, can be expressed in terms of the generator output power, P_g, and the power transmitted to the grid, P_{net}, as

$$V_{DC} = \sqrt{\frac{2}{C} \int (P_g - P_{net}) dt} \tag{6.18}$$

Equation (6.18) calculates the actual value of V_{DC}. The reference value of the active power, $P_{net_{ref}}$, to be transmitted to the network is calculated by comparing the actual DC link voltage, V_{DC}, with the desired DC link voltage reference, $V_{DC_{ref}}$. The error between these two signals is processed by a PI controller, the output of which provides the reference active power $P_{net_{ref}}$, as shown in Figure 6.13. It should be noted that in a physical implementation, the actual value of the DC link voltage, V_{DC}, is obtained from measurements via a transducer.

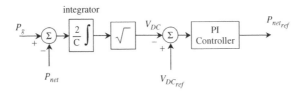

Figure 6.13 Calculation of the active power reference, $P_{net_{ref}}$ (suitable for simulation purposes)

6.1.4.3 Network-side Converter Control and Dynamic Performance Assessment

The objective of the network-side converter controller is to maintain the DC link voltage at the reference value by exporting active power to the network. In addition, the controller is designed to allow the exchange of reactive power between the converter and the network as required by the application specifications.

Load Angle Control Technique
A methodology used to control the network-side converter is also the load angle control technique, where the network-side converter is the sending source, $V_{VSC}\angle\delta$, and the network is the receiving source, $V_{net}\angle 0$. As the network voltage is known, it is selected as the reference, hence the phase angle δ is positive. The inductor coupling these two sources is the reactance X_{net}.

To implement the load angle controller, the reference value of the reactive power, $Q_{net_{ref}}$, may be set to zero for unity power factor operation. Hence the magnitude, V_{VSC}, and angle, δ, required at the terminal of the network-side converter are calculated as

$$\delta = \frac{P_{net_{ref}} X_{net}}{V_{VSC} V_{net}} \tag{6.19}$$

$$V_{VSC} = V_{net} + \frac{Q_{net_{ref}} X_{net}}{V_{VSC}}; \qquad Q_{net_{ref}} = 0 \tag{6.20}$$

From Eqs (6.19) and (6.20), the magnitude of the network-side converter voltage V_{VSC} and angle δ can be obtained as

$$V_{VSC} = g(Q_{net_{ref}}, V_{net}, \delta) = \frac{V_{VSC}\cos\delta + \sqrt{(V_{net}\cos\delta)^2 + 4(Q_{net_{ref}}/3)X_{net}}}{2} \tag{6.21}$$

$$\delta = f(P_{net_{ref}}, V_{net}) = \sin^{-1}\left(\frac{P_{net_{ref}} X_{net}}{3 V_{VSC} V_{net}}\right) \tag{6.22}$$

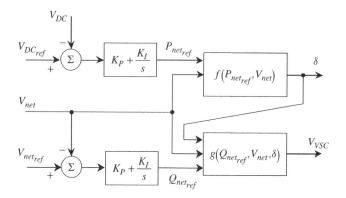

Figure 6.14   Control of active and reactive power by load angle and magnitude control

The second-order quadratic equation Eq. (6.21) needs to be solved to determine the value of V_{VSC}, where only one solution is appropriate. Figure 6.14 shows the control block diagram of the load-angle control methodology. The DC link voltage reference, $V_{DC_{ref}}$, is compared with the actual (or measured) DC voltage, V_{DC}, and the error regulated by a PI controller. The PI controller output $P_{net_{ref}}$ and the reactive power $Q_{net_{ref}}$ are used to find the network-side converter voltage magnitude and angle.

Vector Control Strategy

A block diagram of the vector control of the network-side converter is shown in Figure 6.15. The DC link voltage is maintained by controlling the q axis current and the network terminal voltage is controlled in the d axis. The reference currents are initially determined in the dq frame of the voltage V_{net}, where the voltage vector is aligned with the q axis. Then the reference currents are transformed to the network reference frame and compared with the actual currents. Current error signals are regulated by PI controllers and then decoupling components are added to eliminate the coupling effect between the two axes. Finally, dq components of the voltage V_{net} are added to find the required voltage components at the terminals of the network-side converter in the network reference frame.

6.2   FRC Induction Generator-based (FRC-IG) Wind Turbine

6.2.1   Steady-state Performance

As shown in Figure 1.8, the fully rated converter induction generator-based (FRC-IG) wind turbine is allowed to operate at variable frequency. In order to

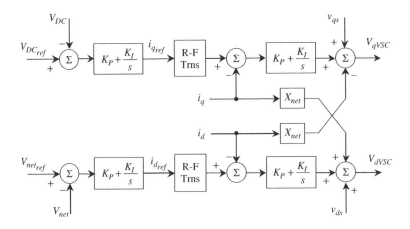

Figure 6.15 Network-side converter control in the dq frame

obtain the steady-state performance of the FRC-IG wind turbine, the machine is represented by the steady-state equivalent circuit given in Chapter 4. However, the reactances were calculated using the machine inductance and the operating frequency. The performance characteristics of the FRC-IG wind turbine for different operating frequencies is shown in Figure 6.16.

In order to follow the maximum power extraction curve shown in Figure 5.8, the generator speed should vary with the wind speed. This is achieved by varying the operating frequency of the induction generator by changing the control signal of the PWM network-side converter. For lower wind speeds, the generator operates at a lower frequency and at higher wind speeds it operates at a higher frequency. As the wind speed varies, the mechanical input, and thus the output power, of the generator increases and, as shown in Figure 6.16b, the reactive power absorbed by the generator remains more or less constant. This requires the slip, which is shown in Figure 6.16a, to be varied with the wind speed. As the maximum speed of operation was limited to 1.2 pu, the upper operating frequency should be limited at 1.2 pu.

6.2.2 Control of the FRC-IG Wind Turbine

The rotor flux oriented control was used in the generator-side converter controller. Figure 6.17 shows the vector diagram representing the operating conditions of an induction generator in a reference frame fixed to the rotor flux (thus $\overline{\psi}_{qr} = 0$). As shown in Figure 6.17, the rotor flux is aligned with the d axis which rotates at the synchronous speed $\overline{\omega}$ (Vas, 1990; Krause *et al.*, 2002).

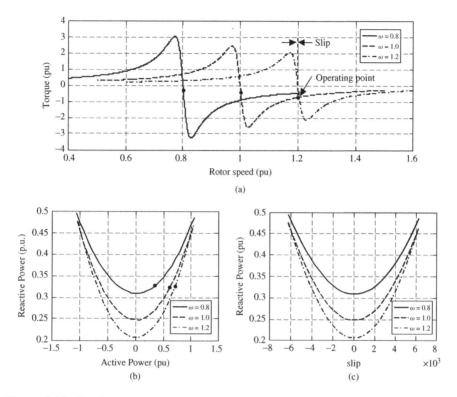

Figure 6.16 Steady-state characteristics of the FRC-IG wind turbine. (a) Torque–speed characteristics; (b) active versus reactive power; (c) slip versus reactive power

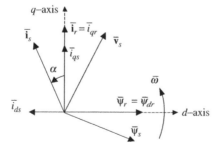

Figure 6.17 Vector diagram representation of the operating conditions of an induction generator in a reference frame fixed to the rotor flux

From Eq. (4.14), $\overline{\psi}_{qr} = \overline{L}_{rr}\overline{i}_{qr} - \overline{L}_m\overline{i}_{qs} = 0$; therefore, the rotor current $\overline{i}_{qr}$ is

$$\overline{i}_{qr} = \frac{\overline{L}_m}{\overline{L}_{rr}}\overline{i}_{qs} \qquad (6.23)$$

Using the expressions for the dq voltages behind a transient reactance $\overline{e}_d$ and $\overline{e}_q$ [from Eqs (4.19) and (4.20)], the electromagnetic torque, $\overline{T}_e$ [given in Eq. (4.38)] is calculated as

$$\overline{T}_e = \frac{\overline{L}_m}{\overline{L}_{rr}}(-\overline{\psi}_{qr}\overline{i}_{ds} + \overline{\psi}_{dr}\overline{i}_{qs}) = \frac{\overline{L}_m}{\overline{L}_{rr}}\overline{\psi}_{dr}\overline{i}_{qs} \qquad (6.24)$$

For an FRC-IG, since the rotor is short circuited, $\overline{v}_{qr} = \overline{v}_{dr} = 0$. Further, if $\overline{\psi}_{qr} = 0$, then $d\overline{\psi}_{qr}/dt = 0$ and if $\overline{\psi}_{dr}$ is a constant then $d\overline{\psi}_{dr}/dt = 0$. Therefore the rotor voltage equations [Eqs (4.9) and (4.10)] can be simplified to

$$\overline{v}_{dr} = \overline{R}_r\overline{i}_{dr} = 0 \qquad (6.25)$$

$$\overline{v}_{qr} = \overline{R}_r\overline{i}_{qr} + s\overline{\omega}\overline{\psi}_{dr} = 0 \qquad (6.26)$$

From Eq. (6.26), the slip speed can be obtained as

$$s\overline{\omega} = -\frac{\overline{R}_r\overline{i}_{qr}}{\overline{\psi}_{dr}} \qquad (6.27)$$

Substituting Eq. (6.25) into Eq. (4.13), $\overline{\psi}_{dr}$ is obtained as

$$\overline{\psi}_{dr} = -\overline{L}_m\overline{i}_{ds} \qquad (6.28)$$

Substituting Eq. (6.28) into Eqs (6.24) and (6.27), the electromagnetic torque and slip speed can be rewritten as

$$\overline{T}_e = -\frac{\overline{L}_m^2}{\overline{L}_{rr}}\overline{i}_{ds}\overline{i}_{qs} \qquad (6.29)$$

$$s\overline{\omega} = \frac{\overline{R}_r\ \overline{i}_{qr}}{\overline{L}_m\ \overline{i}_{ds}} \qquad (6.30)$$

Substituting Eq. (6.23) into Eq. (6.30):

$$s\overline{\omega} = \frac{\overline{R}_r\ \overline{i}_{qs}}{\overline{L}_{rr}\ \overline{i}_{ds}} \qquad (6.31)$$

With $\bar{i}_{dr} = 0$, Eq. (4.11) reduces to

$$\overline{\psi}_{ds} = -\overline{L}_{ss}\bar{i}_{ds} \tag{6.32}$$

Substituting Eq. (6.23) into Eq. (4.12), the following equation was obtained:

$$\overline{\psi}_{qs} = -\overline{L}_{ss}\bar{i}_{qs} + \overline{L}_{m}\frac{\overline{L}_{m}}{\overline{L}_{rr}}\bar{i}_{qs} = -\left(\overline{L}_{ss} - \frac{\overline{L}_{m}^{2}}{\overline{L}_{rr}}\right)\bar{i}_{qs} = -\overline{L}'\bar{i}_{qs} \tag{6.33}$$

where $L' = \overline{L}_{ss} - (\overline{L}_{m}^{2}/\overline{L}_{rr})$. Substituting $\overline{\psi}_{ds}$ and $\overline{\psi}_{qs}$ from Eqs (6.32) and (6.33) into Eqs (4.7) and (4.8), the stator voltages in the steady state are given by

$$\overline{v}_{ds} = -\overline{R}_{s}\bar{i}_{ds} + \overline{\omega}\overline{L}'\bar{i}_{qs} \tag{6.34}$$

$$\overline{v}_{qs} = -\overline{R}_{s}\bar{i}_{qs} - \overline{\omega}\overline{L}_{ss}\bar{i}_{ds} \tag{6.35}$$

The stator voltage $\overline{v}_{ds}$ includes the voltage $\overline{\omega}\overline{L}'\bar{i}_{qs}$ and $\overline{v}_{qs}$ includes the voltage $-\overline{\omega}\overline{L}_{ss}\bar{i}_{ds}$. These terms give the cross-coupling of the dq axes voltages with qd axes currents. It follows that the d axis stator voltage is also affected by the q axis stator current and the q axis stator voltage is also affected by the d axis stator current. To eliminate the coupling effect, $\overline{\omega}\overline{L}'\bar{i}_{qs}$ and $-\overline{\omega}\overline{L}_{ss}\bar{i}_{ds}$ are added in the control system. Then $\bar{i}_{ds}$ is controlled through $\overline{v}_{ds}$ and $\bar{i}_{qs}$ is controlled through $\overline{v}_{qs}$ independently. The flux and torque control loops of the generator-side converter controller are shown in Figure 6.18.

In the flux control loop, the reference d axis stator current i_{ds}^{ref} sets the air-gap flux level. The reference d axis current is compared with its actual value and the error signal is regulated by the PI controller. The PI controller output and the decoupling term are added to obtain the d axis stator voltage. The reference q axis stator current is obtained using the generator torque-speed curve defined in Figure 5.8 and Eq. (6.29). This is compared with its actual value and the error signal is regulated by the PI controller. The output of the controller is added to the decoupling term to determine the q axis stator voltage.

As shown in Figure 6.18, the generator terminal frequency was controlled by adding the rotor speed and the slip speed given in Eq. (6.31).

The ratio of the stator reactance to the stator resistance for a larger induction machine is much higher than that of a smaller induction machine (Krause et al., 2002). For example, this ratio is about 20 for a 2 MW induction generator that is employed for a wind turbine (in contrast to a ratio of 2 for a 3 HP

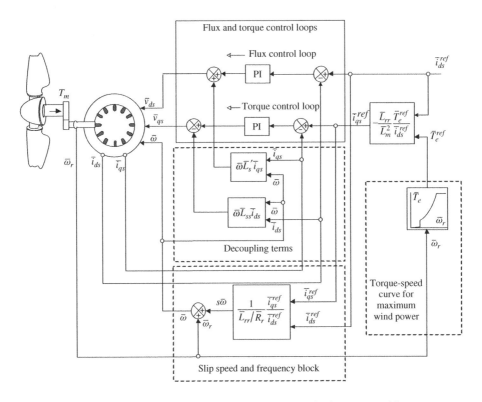

Figure 6.18   Block diagram of rotor flux oriented control of generator-side converter

induction machine). Therefore, the PI controllers defined by Eqs. (6.34) and (6.35) can be simplified to

$$\overline{v}_{ds} = \overline{\omega}\overline{L}'\,\overline{i}_{qs} \tag{6.36}$$

$$\overline{v}_{qs} = -\overline{\omega}\overline{L}_{ss}\overline{i}_{ds} \tag{6.37}$$

Hence two PI controllers can be simplified as shown in Figure 6.19 without the decoupling terms.

The network-side controller was controlled as described in Section 6.1.4.3.

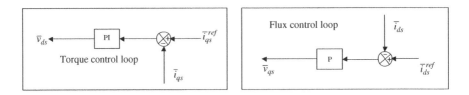

Figure 6.19   Simplified generator controller

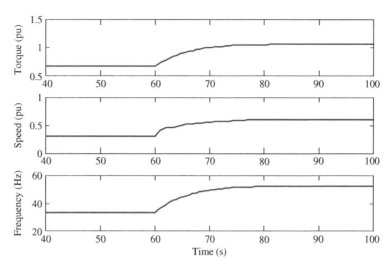

Figure 6.20   FRC-IG responses for a 100% increase and decrease in the mechanical torque input

6.2.3   Performance Characteristics of the FRC-IG Wind Turbine

The behaviour of the FRC-IG wind turbine was explored with step changes in the mechanical input torque. In this simulation, the mechanical input torque was increased from 0.3 to 0.6 pu at $t = 60$ s. Figure 6.20 illustrates the electrical torque, rotor speed and frequency of the FRC-IG. As shown, the electrical torque output of the FRC-IG follows the new torque reference after a short transient period.

References

Akhmatov, V., Nielsen, A. H. and Pedersen, J. K. (2003) Variable-speed wind turbines with multi-pole synchronous permanent magnet generators. Part I. Modelling in dynamic simulation tools, *Wind Engineering*, **27**, 531–548.

Chen, Z. and Spooner, E. (1998) Grid interface options for variable-speed permanent-magnet generators, *IEE Proceedings Electric Power Application*, **145** (4), 273–283.

Fox, B., Flynn, D., Bryans, L., Jenkins, N., Milborrow, D., O'Malley, M., Watson, R. and Anaya-Lara, O. (2007) *Wind Power Integration: Connection and System Operational Aspects*, IET Power and Energy Series 50, Institution of Engineering and Technology, Stevenage, ISBN 10: 0863414494.

Grauers, A. (1996) Design of direct driven permanent magnet generators for wind turbines, PhD Thesis, Chalmers University of Technology, Rep. No. 292 L.

Krause, P. C., Wasynczuk, O. and Shudhoff, S. D. (2002) *Analysis of Electric Machinery and Drive Systems*, 2nd edn, Wiley-IEEE Press, New York.

Kundur, P. (1994) *Power System Stability and Control*, McGraw-Hill, New York, ISBN 0-07-035958-X.

Vas, P. (1990) *Vector Control of AC Machines*, Oxford University Press, New York.

7

Influence of Rotor Dynamics on Wind Turbine Operation

New designs of wind turbines continue to increase in rotor size in order to extract more power from wind. As the rotor diameters increase, the flexibility of the rotor structure increases as well as the influence of the mechanical drive train on the electrical performance of the wind turbine. When the length of the rotor blades increases, the frequencies of the torque oscillations reduce and these oscillations may then interact with the low-frequency modes of the electrical network.

In smaller FSIG wind turbines, the induction generator acts as an effective damper, which helps to reduce the magnitude of the torque oscillations. However, it has been reported that these oscillations are still significant and must be taken into account when analysing the dynamic performance of FSIG wind turbines for transient stability (Akhmatov, 2003).

In variable-speed DFIG wind turbines, which operate at a defined torque, the damping contribution of the generator is low because the torque no longer varies rapidly as a function of the rotor speed. Also, active damping techniques are often used to stabilize the mechanical systems of large variable-speed wind turbines (Burton *et al.*, 2001). Recently, power system stabilizers (PSSs) have been proposed to enable DFIG wind turbines to contribute positively to network damping (Hughes *et al.*, 2005). If any of the frequencies of mechanical vibration of the rotor structure lies within the bandwidth of the PSS, then resonance or adverse control loop interactions may arise, which will affect the performance of both the mechanical and electrical systems of the wind turbine.

Wind Energy Generation: Modelling and Control   Olimpo Anaya-Lara, Nick Jenkins, Janaka Ekanayake, Phill Cartwright and Mike Hughes
© 2009 John Wiley & Sons, Ltd

A number of authors have addressed the representation of wind turbines in large power system studies. Papathanassiou and Papadopoulos (1999), Akhmatov (2002) and Ackermann (2005) discussed a two-mass model which takes into account the shaft flexibility but neglects the dynamics of the blades. In the work of Papathanassiou and Papadopoulos (2001), the blade dynamics are included by representing the individual blades by masses connected to the hub via springs. However, the order of their model is high as it consists of six masses and five springs.

High-order representations of the structural dynamics may not be suitable for large power system simulation studies. An alternative approach is to use a reduced multi-mass model to represent the rotor structural dynamics with appropriate accuracy. A three-mass model may be used to represent shaft and blade dynamics, but this can be simplified further into an effective two-mass model that gives an accurate representation of the dominant lower frequency component of the rotor structure (Ramtharan, 2008). It should be noted that from the power system viewpoint the lowest frequency mode (associated with the blades) is of most relevance and hence an effective two-mass model incorporating this mode is then appropriate.

7.1 Blade Bending Dynamics

The torsional flexibility of the shaft and the bending flexibility of the blades both contribute to the wind turbine torque oscillations. The torsional oscillations of the shaft can be represented using a simple model as explained by Papathanassiou and Papadopoulos (1999), Akhmatov (2002) and Ackermann (2005). However, the representation of the blades is not straightforward due to the non-uniform distribution of their mass, stiffness and twist angle. For simplicity, these physical properties can be assumed to be uniform in order to analyse the dominant vibration mode of the blades (Eggleston and Stoddard, 1987). If a more accurate representation of the dynamic properties of the blades is required, then finite element techniques may be employed (Géradin and Rixen, 1997; Bossanyi, 2003).

The bending modes of a blade are defined in two orthogonal planes: (i) out-of-plane, which describes the motion of the blade perpendicular to the rotor plane and (ii) in-plane, which describes the motion of the blade in the rotor plane (Johnson, 1980). As the motion of the out-of-plane modes is normal to the direction of rotation of the rotor, it does not directly couple to the drive train and therefore it is not necessary to include it in the representation of the drive train dynamics. However, some of the in-plane modes directly couple to the drive train.

In the in-plane bending mode, there are two asymmetric modes and one symmetric mode of vibration. The asymmetric modes do not couple with the drive train and therefore may be neglected. In the symmetric mode, all the blades oscillate in unison with one another with respect to the hub (Figure 7.1a).

7.2 Derivation of Three-mass Model

This is illustrated in Figure 7.1, where the blade bending dynamics illustrated in Figure 7.1a are represented as a simple torsional system as shown in Figure 7.1b. Since the blade bending occurs at a distance from the joint between the blade and the hub, the blade can be split into two parts, OA and AB. The rigid blade sections OA1, OA2 and OA3 are collected into the hub and have an inertia J_2 and the rest of the blade sections A1B1, A2B2 and A3B3 are collected as a ring flywheel with inertia J_1 about the shaft. The inertias J_1 and J_2 are connected via three springs, which represent the flexibility of individual blades.

By simplifying the rotor mode to two inertias connected via springs (Figure 7.1b), the drive train of the turbine can be represented by a three-mass model as shown in Figure 7.2, where J_1 represents the inertia of the flexible blade section, J_2 represents the combined inertia of the hub and the rigid blade section, J_3 is the generator inertia, K_1 is the effective blade stiffness and K_2 represents the shaft stiffness (resultant stiffness of both the low- and high-speed shafts). The generator inertia J_3, the shaft stiffness K_2 and the

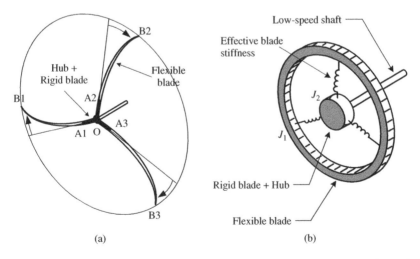

(a)   (b)

Figure 7.1 Equivalent blade inertia and stiffness of the in-plane rotor symmetric mode. (a) in-plane rotor symmetric bending mode; (b) equivalent torsional representation

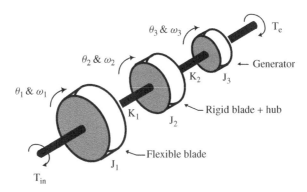

Figure 7.2   Three-mass model of drive train including blade and shaft flexibilities

rotor total inertia $J_1 + J_2$ are known variables. Therefore, two more equations are necessary to determine all five parameters of this three-mass model.

The equations describing the two- and three-mass models are given in Table 7.1 (Tse *et al.*, 1963; Thomson, 1993; Harris, 1996). The two frequencies of vibration in Eq. (7.4), f_1 and f_2, can be obtained by conducting a spectral analysis of the low-speed shaft torque (through simulation) or by physical measurements.

7.2.1   Example: 300 kW FSIG Wind Turbine

In this example, the three-mass model for a 300 kW FSIG wind turbine is calculated (the wind turbine data are given in Appendix D). The known parameters of the rotor are $J_3 = 0.102 \times 10^6$ kgm^2, $J_1 + J_3 = 0.129 \times 10^6$ kgm^2 and $K_2 = 5.6 \times 10^7$ N m rad^{-1}. All these parameters are referred to the low-speed shaft. A time domain simulation was carried out using the Garrad Hassan wind turbine simulation program GH BLADED.

In order to identify the frequencies of vibration of the rotor, the rotor was excited by applying a voltage sag of 20% (80% retained voltage) to the generator at 5 s with a duration of 200 ms. Figure 7.3a shows the response of the low-speed shaft torque during the voltage sag and the corresponding frequency spectrum is given in Figure 7.3b. Substituting the spectral peak frequencies shown in Figure 7.3b in Eq. (7.4) gives the three-mass model parameters: $J_1 = 0.111 \times 10^6$, $J_2 = 0.018 \times 10^6$, $J_3 = 0.102 \times 10^6$, $K_1 = 2.1 \times 10^7$ and $K_2 = 5.6 \times 10^7$. From this example, it can be noticed that the effective flexibility of the blade, K_1, is 0.4 times smaller than the flexibility of the shaft, K_2, that is, the blade flexibility is much more important than that of the shaft.

Table 7.1 Equations of the two- and three-mass models used to represent the rotor structural dynamics

Two-mass model	Three-mass model
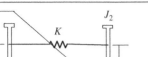	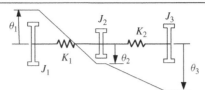

Dynamic equations: **Dynamic equations:**

$$J_1 \frac{d^2}{dt^2}\theta_1 = -K(\theta_1 - \theta_2) \tag{7.1}$$
$$J_2 \frac{d^2}{dt^2}\theta_2 = -K(\theta_2 - \theta_1)$$

$$J_1 \frac{d^2}{dt^2}\theta_1 = -K_1(\theta_1 - \theta_2)$$
$$J_2 \frac{d^2}{dt^2}\theta_2 = -K_2(\theta_2 - \theta_1) - K_2(\theta_2 - \theta_3) \tag{7.2}$$
$$J_3 \frac{d^2}{dt^2}\theta_3 = -K_2(\theta_3 - \theta_2)$$

Natural frequency of vibration: **Natural frequencies of vibration:**

$$f = \frac{1}{2\pi} \sqrt{\frac{K}{\left(\frac{1}{J_1} + \frac{1}{J_2}\right)^{-1}}} \tag{7.3}$$

$$f_1 = \frac{1}{2\pi}\left(-\frac{b}{2} - \frac{\sqrt{b^2 - 4c}}{2}\right)^{\frac{1}{2}}$$
$$f_2 = \frac{1}{2\pi}\left(-\frac{b}{2} + \frac{\sqrt{b^2 - 4c}}{2}\right)^{\frac{1}{2}} \tag{7.4}$$

Magnitude of oscillation: **Magnitudes of oscillations:**

$$\frac{\theta_1}{\theta_2} = -\frac{J_1}{J_2} \tag{7.5}$$

$$\frac{\theta_1}{\theta_2} = \frac{K_1}{(K_1 - J_1\omega^2)}$$
$$\frac{\theta_2}{\theta_3} = \frac{(K_2 - J_3\omega^2)}{K_2} \tag{7.6}$$

$$b = -\left[K_1\left(\frac{1}{J_1} + \frac{1}{J_2}\right) + K_2\left(\frac{1}{J_2} + \frac{1}{J_3}\right)\right]; \quad c = K_1 K_2 \frac{J_1 + J_2 + J_3}{J_1 J_2 J_3}$$

However, representation of both shaft and blade flexibilities increases the order of the model, which may not be desirable in large power system studies. In addition, for power system dynamic studies the frequency that is likely to be most significant is the lowest frequency component, which in this particular example is 2.7 Hz. Hence an effective two-mass model, which takes into

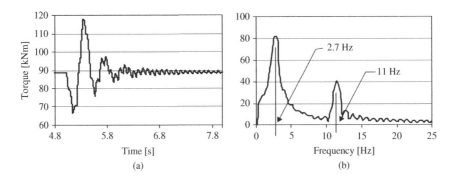

Figure 7.3 300 kW FSIG-based wind turbine. Variation of the low-speed shaft torque and its harmonic spectrum during a 20% terminal voltage drop. (a) Low-speed shaft torque variation; (b) FFT of low-speed shaft torque

account both shaft and blades flexibilities but that only represents the dominant low-frequency component of the rotor structure, is derived.

7.3 Effective Two-mass Model

The three-mass model shown in Figure 7.2 has two coupled modes with frequencies f_1 and f_2. If the blade flexibility is neglected (making $K_1 = \infty$), a two-mass model would have a natural frequency of vibration f_{shaft} as illustrated schematically by the frequency spectrum in Figure 7.4.

Similarly, if the shaft flexibility is neglected (making $K_2 = \infty$), the two-mass model would have a natural frequency of vibration f_{blade}. As can be seen in Figure 7.4, neither of these assumptions gives the true dominant frequency of vibration of the rotor dynamics, f_1, as obtained from the more detailed three-mass model. The frequency component f_2 (11.0 Hz) in the

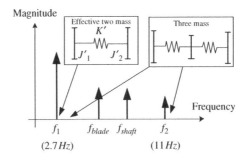

Figure 7.4 Frequency components of multi-mass system: f_1 and f_2, frequency components of coupled modes in the three-mass model; f_{shaft}, two-mass model with shaft flexibility only; f_{blade}, two-mass model with blade flexibility only

three-mass model is higher than the low-frequency modes of oscillation in the electrical system (Kundur, 1994) and so it may be neglected for many power system studies. Hence an effective two-mass model can be derived which represents only the lower frequency component f_1 (2.7 Hz) of the three-mass model.

Using again the example given for the FSIG wind turbine, three equations are required to find the parameters J_1', J_2' and K'. These equations are obtained as follows.

The total moment of inertia of the turbine is given as

$$J_1' + J_2' = J_{\text{total}} = 0.231 \times 10^6 \text{ kg m}^2 \tag{7.7}$$

The lowest frequency component of the rotor was 2.7 Hz; considering the natural frequency of vibration of the two-mass model in Table 7.1 [Eq (7.3)] gives

$$f_1 = \frac{1}{2\pi} \sqrt{\frac{K'}{(1/J_1' + 1/J_2')^{-1}}} = 2.7 \text{ Hz} \tag{7.8}$$

For the final equation, the magnitude ratio of the oscillation of the effective two- and the three-mass models [from Eqs (7.5) and (7.6) in Table 7.1] are used:

$$\frac{\theta_1'}{\theta_2'} = -\frac{J_1'}{J_2'} = \frac{\theta_1}{\theta_3} = \frac{K_1(K_2 - J_3\,\omega^2)}{K_2(K_1 - J_1\,\omega^2)} = -0.92 \tag{7.9}$$

Solving Eqs (7.7)–(7.9) gives the effective two-mass model parameters as $J_1' = 0.111 \times 10^6$, $J_2' = 0.12 \times 10^6$ and $K' = 1.66 \times 10^7$. The effective two-mass model was represented in GH BLADED by making the rotor blades rigid and the shaft stiffness was changed from 5.6×10^7 to 1.66×10^7 N m rad^{-1} (from the actual shaft stiffness to the effective two-mass model shaft stiffness).

Figure 7.5 shows the responses of the low-speed shaft torque of the 300 kW FSIG wind turbine during a voltage sag of 20% (80% retained voltage), with full representation of the rotor dynamics and with the effective two-mass model. For this study, the generator rotor electrical transients were included and the stator transients were neglected. The voltage disturbance was applied at the generator terminals at 5 s with a duration of 200 ms.

It can be seen in Figure 7.5 that the effective two-mass model representation gives a similar response (if only the lowest frequency component is considered) to that of the model with full representation of the rotor structural dynamics.

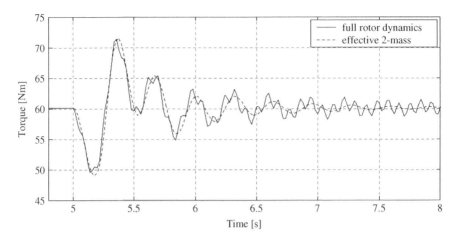

Figure 7.5 Variation of the low-speed shaft torque with full rotor dynamic representation and effective two-mass model

In order to see the performance of the two-mass model with only the shaft dynamics, the shaft stiffness was brought back to 5.6×10^7 from 1.66×10^7 N m rad^{-1} (from effective two-mass model stiffness to actual shaft stiffness) while keeping the blades rigid. The low-speed shaft response of the FSIG for this case is shown in Figure 7.6. The two-mass model (considering only the shaft flexibility) represents the frequency component f_{shaft} instead of the dominant frequency f_1 (see Figure 7.4). The response of a single-mass model is also shown in Figure 7.6. It can be seen that the response of a two-mass model considering only shaft flexibility is closer to the single-mass model than to the model with full rotor dynamics representation.

7.4   Assessment of FSIG and DFIG Wind Turbine Performance

The performance of FSIG and DFIG wind turbines during electrical faults in the network was assessed using the following model representations of the rotor structural dynamics: (i) single-mass model, (ii) two-mass model with only shaft flexibility while keeping the blades rigid, (iii) effective two-mass model to represent the lowest frequency of vibration of the rotor structure and (iv) full representation of the rotor structural dynamics.

The responses obtained for the FSIG wind turbine are shown in Figures 7.7 and 7.8. An electrical fault is applied at 5 s with a duration of 200 ms. Due to the fault, the voltage at the terminals of the wind turbine drops to 80% of the nominal voltage (20% retained voltage). The low-speed shaft torque response is shown in Figure 7.7 and the rotor current response is given in Figure 7.8. It

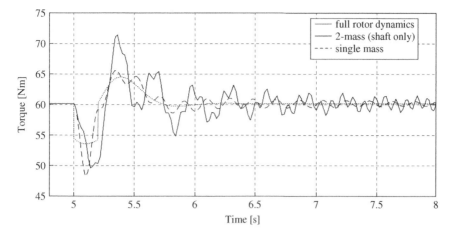

Figure 7.6   Variation of the low-speed shaft torque with full rotor dynamic representation, two-mass model with only shaft dynamics and a single-mass model

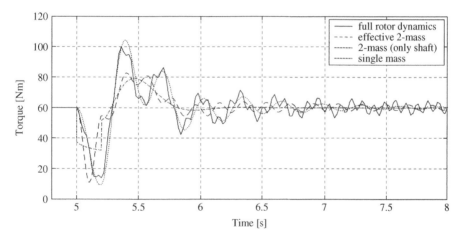

Figure 7.7   Variation of low-speed shaft torque of 300 kW FSIG during an 80% voltage drop (20% retained voltage) due to a fault in the network, with different model representations of the rotor structural dynamics

can be observed that the low-frequency component obtained with the effective two-mass model is very similar to that obtained with the full representation of the rotor structure. It can also be seen that the two-mass model considering only the shaft flexibility does not agree well with the actual response of the turbine.

In the DFIG case, the wind turbine was controlled using the current-mode control strategy described by Ekanayake *et al.* (2003) and Holdsworth *et al.*

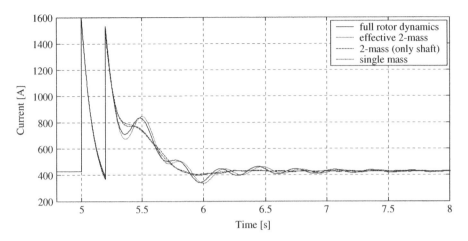

Figure 7.8 Variation of rotor current of 300 kW FSIG during an 80% voltage drop (20% retained voltage) due to a fault in the network, with different model representations of the rotor structural dynamics

(2003) to extract maximum power from wind (for the prevailing wind velocity) while operating at unity power factor. A crowbar protection system was implemented to protect the DFIG converter during the fault. For the studies presented, normal converter operation takes place for rotor current magnitudes less than 2.5 kA (1.5 rated value). For rotor currents greater than 2.5 kA, the crowbar acts to short-circuit the rotor through an external resistor.

The studies were conducted by applying a three-phase fault (with a clearance time of 200 ms), which caused a voltage drop at the DFIG terminals of 85% (15% retained voltage). As in the FSIG case, different models were used to represent the rotor structural dynamics of the DFIG wind turbine. The responses obtained for a 2 MW DFIG wind turbine are shown in Figures 7.9 and 7.10. Figure 7.9 shows the variation of the low-speed shaft torque and Figure 7.10 shows the variation of the rotor current of the DFIG where the satisfactory operation of the crowbar protection can also be observed.

The representation of the DFIG with the effective two-mass model gives the lowest frequency component of the rotor structure as expected. It can be seen that the response obtained with the two-mass model considering only the shaft flexibility differs significantly from the actual response of the turbine.

The fault studies conducted with FSIG and DFIG wind turbines show that the torque oscillations of a two-mass model with only the shaft flexibilities represented are more benign than those obtained with the effective two-mass model. The two-mass model with representation of only shaft flexibilities may

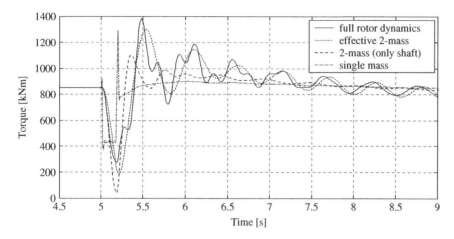

Figure 7.9   Variation of the low-speed shaft torque of 2 MW DFIG during a fault (85% voltage drop).

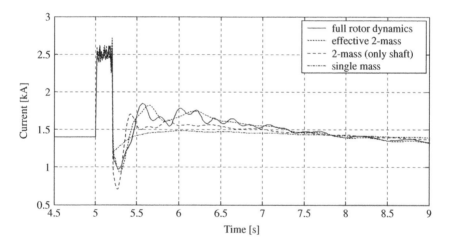

Figure 7.10   Variation of rotor current of 2 MW DFIG during a fault (85% voltage drop).

therefore not be appropriate for power system transient stability studies as some torque oscillations, which may interact with the electrical system, are not fully taken into account.

For DFIG control design, the three-mass model may be useful to help identify possible control loop interaction problems. Once a satisfactory control scheme has been designed, then a reduced order model can be employed for power system analysis depending on the specific dynamics of interest.

Acknowledgement

The material in this chapter is based on the PhD thesis "Control of variable speed wind turbine generators", University of Manchester, 2008 by Gnanasambandapillai Ramtharan, and used with his permission.

References

Ackermann, T. (2005) *Wind Power in Power Systems*, John Wiley & Sons, Ltd, Chichester, pp. 536–546, ISBN: 0-470-85508-8.

Akhmatov, V. (2002) Variable-speed wind turbines with doubly-fed induction generators. Part I. Modelling in dynamic simulation tools, *Wind Engineering*, **26** (2), 85–108.

Akhmatov, V. (2003) Analysis of dynamic behaviour of electric power systems with large amount of wind power. PhD Thesis. Technical University of Denmark, pp. 30–38; http://server.oersted.dtu.dk/eltek/res/phd/00–05/20030430-va.html.

Bossanyi, E. A. (2003) *GH Bladed Theory and User Manuals*, Garrad Hassan and Partners Limited, Bristol, Document No. 282/BR/009, Issue No. 12.

Burton, T., Sharpe, D., Jenkins, N. and Bossanyi, E. (2001) *Wind Energy Handbook*, John Wiley & Sons Ltd, Chichester, ISBN 0-471-48997-2, pp. 488–489.

Eggleston, M. and Stoddard, S. (1987) *Wind Turbine Engineering Design*, Van Nostrand Reinhold, New York, ISBN 0-442-22195-9.

Ekanayake, J. B., Holdsworth, L., Wu, X. and Jenkins, N. (2003) Dynamic modelling of doubly fed induction generator wind turbines, *IEEE Transactions on Power Systems*, **18** (2), 803–809.

Géradin, M. and Rixen, D. (1997) *Mechanical Vibrations: Theory and Application to Structural Dynamics*, 2nd edn, John Wiley & Sons, Inc., New York, ISBN 0-471-97546-5.

Harris, C. M. (1996) *Shock and Vibration Handbook*, 4th edn, McGraw-Hill, New York, pp. 38.1–38.14, ISBN 0-07-026920-3.

Holdsworth, L., Wu, X., Ekanayake, J. B. and Jenkins, N. (2003) Comparison of fixed speed and doubly fed induction wind turbines during power system disturbances, *IEE Proceedings Generation Transmission Distribution*, **150** (3), 343–352.

Hughes, F. M., Anaya-Lara, O., Jenkins, N. and Strbac, G. (2005) Control of DFIG-based wind generation for power network support, *IEEE Transactions on Power Systems*, **20** (4), 1958–1966.

Johnson, W. (1980) *Helicopter Theory*, Dover Publications, New York, pp. 381–403, ISBN 0-86-68230-7.

Kundur, P. (1994) *Power System Stability and Control*, McGraw-Hill, New York, ISBN 0-07-035958-X.

Papathanassiou, S. A. and Papadopoulos, M. P. (1999) Dynamic behaviour of variable speed wind turbines under stochastic wind, *IEEE Transactions on Energy Conversion*, **14** (4), 1617–1623.

Papathanassiou, S. A. and Papadopoulos, M. P. (2001) Mechanical stress in fixed speed wind turbines due to network disturbance, *IEEE Transactions on Energy Conversion*, **16** (4), 361–363.

Ramtharan, G. (2008) *Control of variable speed wind turbine generators*, PhD Thesis, University of Manchester, UK

Thomson, W. T. (1993) *Theory of Vibration with Applications*, 4th edn, Chapman & Hall, London, pp. 131–145, 268–337, ISBN 0-412-78390-8.

Tse, F. S., Morse, I. E. and Hinkle, R. T. (1963) *Mechanical Vibrations*, Prentice Hall, Englewood Cliffs, NJ, pp. 95–115.

8

Influence of Wind Farms on Network Dynamic Performance

A power network is rarely in a steady operating condition, with the load varying as both industrial and domestic consumers switch equipment on and off and, with renewables, a further variable component is introduced on the generation side.

Consumer load and renewable generation vary on a seasonal and daily basis and are influenced by weather conditions and a host of other things. Predictable variations can be accommodated by forecasting demand and generation capability to ensure that appropriate levels of generation are available. The unpredictable elements need to be accommodated by the controllers of the network and the generators. In addition, severe disturbances can result due to equipment failure and due to faults on the transmission and distribution networks. Since a power network is continually being subjected to disturbances, it is essential that it can accommodate these disturbances and can operate in a stable manner over the required range of operation and maintain the expected quality of supply to the consumers (Vittal, 2000).

Power system stability has many aspects, but attention will be confined here to dynamic stability, transient stability and voltage stability considerations (IEEE/CIGRE, 2004).

8.1 Dynamic Stability and its Assessment

In the power systems context, dynamic stability refers to the ability of a power network to maintain an operating condition in the presence of small disturbances. If a small disturbance results in conditions moving irrevocably

Wind Energy Generation: Modelling and Control Olimpo Anaya-Lara, Nick Jenkins,
Janaka Ekanayake, Phill Cartwright and Mike Hughes
© 2009 John Wiley & Sons, Ltd

away from the original operating point, then the system is classed as being dynamically unstable (Kundur, 1994).

Although a power network is a complex, nonlinear system, by confining attention to small variations about a particular steady-state operating point, the nonlinear equations representing the system can be linearized. Once the system model is available in linearized form, dynamic stability can be assessed using any of the well-developed analysis techniques of linear algebra.

The most widely used approach is eigenvalue analysis (Wong *et al*, 1988; Grund *et al*, 1993). The system equations are arranged in state–space form and the eigenvalues are calculated from the system state matrix. If all of the eigenvalues lie in the left half of the complex plane, then the system is stable. If any of the eigenvalues lie in the right half of the complex plane, then the system is unstable.

Purely real eigenvalues denote aperiodic modes, i.e. modes that grow or decay exponentially with time. For stability, all modes must decay so that all real eigenvalues must be negative and reside in the left half of the complex plane. Complex eigenvalues occur in complex conjugate pairs and indicate an oscillatory mode of behaviour. The value of the imaginary part indicates the frequency of oscillation and the real part gives the exponential growth rate of the magnitude of the oscillations. For the oscillations to decay as time progresses, the real part of a complex eigenvalue must be negative. Hence, for stability all eigenvalues need to have negative real parts and hence reside in the left half of the complex plane.

A very simple and brief treatment of state–space modelling, linearization of nonlinear state equations and eigenvalue analysis is provided in Appendices A–C.

8.2 Dynamic Characteristics of Synchronous Generation

The major dynamic features that constrain the operational capabilities of a power network are dictated by the requirement that all the synchronous generators directly connected to the network must operate in synchronism with each other. Hence network stability limitations are mainly imposed by the interaction characteristics of its synchronous generation. The generators of wind farms, whether FSIG, DFIG or FRC based, operate asynchronously with respect to the network frequency, so that although they do influence the behaviour of mixed generation networks, stability considerations remain essentially linked with the synchronous generation of the network. As a consequence of this, prior to looking into the influence of wind generation on dynamic stability and performance, it is important to give some coverage of the basic dynamic characteristics of conventional synchronous generation and how these impact on network dynamic behaviour.

8.3   A Synchronizing Power and Damping Power Model of a Synchronous Generator

A simple model of a synchronous generator that can usefully be employed in the study and explanation of generator dynamic behaviour under oscillatory network operating conditions is one based on the concept of synchronizing power and damping power (DeMello and Concordia, 1969). Under oscillatory conditions, the model simply relates the variations in the power output of the generator to the variations in its rotor angle at a specified frequency of concern. The power output response, ΔP_e, is separated into two components, one in phase with oscillations in the rotor angle, $\Delta \delta_r$ and the other in phase with oscillations in the rotor speed, $\Delta \omega_r$.

Let the transfer function, $g_g(s)$, represent the dynamic relationship between rotor angle variations and electrical power variations of the generator, taking into account the control loops of the generator and the influence of the network load. Then,

$$\Delta P_e(s) = g_g(s)\Delta \delta_r(s) \tag{8.1}$$

For oscillations in rotor angle of frequency, ω_{osc}, the above can be re-expressed as

$$\Delta P_e(j\omega_{osc}) = g_g(j\omega_{osc})\Delta \delta_r(j\omega_{osc})$$
$$= (R_e + jX_e)\Delta \delta_r(j\omega_{osc}) \tag{8.2}$$

where R_e is the real part of $g_g(j\omega_{osc})$ and jX_e is its imaginary part.

Since $\Delta \omega_r = d(\Delta \delta_r)/dt$ for $\Delta \delta_r$ in radians and $\Delta \omega_r$ in radians per second, for sinusoidal oscillations in rotor angle defined by $\Delta \delta_r = \delta_m \sin(\omega_{osc}t)$

$$\frac{d}{dt}\Delta \delta_r = \omega_{osc}\delta_m \cos(\omega_{osc}t) = \omega_{osc}\delta_m \sin\left(\omega_{osc}t + \frac{\pi}{2}\right) \tag{8.3}$$

Hence rotor speed, $\Delta \omega_r$, is phase displaced by $\pi/2$ rad from rotor angle, $\Delta \delta_r$, so that in vectorial form

$$\Delta \omega_r = j\omega_{osc}\Delta \delta_r \tag{8.4}$$

From Eqs (8.2) and (8.4), the expression for the power–angle relationship can be rewritten (ignoring the operator $j\omega_{\mathrm{osc}}$ for convenience) as

$$\Delta P_{\mathrm{e}} = C_{\mathrm{s}}\Delta\delta_{\mathrm{r}} + jC_{\mathrm{d}}\Delta\omega_{\mathrm{r}} = \Delta P_{\mathrm{es}} + j\Delta P_{\mathrm{ed}} \tag{8.5}$$

where $\Delta P_{\mathrm{es}} = C_{\mathrm{s}}\Delta\delta_{\mathrm{r}}$ and $\Delta P_{\mathrm{ed}} = C_{\mathrm{d}}\Delta\omega_{\mathrm{r}}$.

The power response has therefore been split into two components, a synchronizing power component, ΔP_{es}, that is in phase with rotor angle oscillations and a damping power component, ΔP_{ed}, that is in phase with rotor speed oscillations.

$C_{\mathrm{s}}(= R_{\mathrm{e}})$ is the synchronizing power coefficient and $C_{\mathrm{d}}(= X_{\mathrm{e}}/\omega_{\mathrm{osc}})$ for $\Delta\omega_{\mathrm{r}}$ in radians per second [or by $C_{\mathrm{d}}(= 2\pi f X_{\mathrm{e}}/\omega_{\mathrm{osc}})$ for $\Delta\omega_{\mathrm{r}}$ in per unit (pu) speed] is the damping power coefficient.

Hence, for the study of generator behaviour under oscillatory conditions, the generator inertia relationships can be expressed as

$$\frac{d}{dt}\Delta\omega_{\mathrm{r}} = \frac{1}{2H}(\Delta P_{\mathrm{m}} - \Delta P_{\mathrm{e}}) \tag{8.6}$$

$$\frac{d}{dt}\Delta\delta_{\mathrm{r}} = 2\pi f \Delta\omega_{\mathrm{r}} \tag{8.7}$$

(with $\Delta\delta_{\mathrm{r}}$ in radians and $\Delta\omega_{\mathrm{r}}$ in pu speed), where $\Delta P_{\mathrm{e}} = C_{\mathrm{s}}\Delta\delta_{\mathrm{r}} + jC_{\mathrm{d}}\Delta\omega_{\mathrm{r}}$.

The synchronizing and damping power model is shown in block diagram form in Figure 8.1.

The block diagram provides a very simple second-order equivalent representation of the generator behaviour at the oscillation frequency of concern,

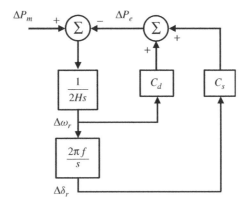

Figure 8.1 Simplified synchronizing and damping torque model of a synchronous generator

ω_{osc}, and gives rise to the following relationship:

$$\Delta\delta = \frac{\pi f}{H} \frac{1}{\left(s^2 + \frac{C_d}{2H}s + \frac{\pi f C_s}{H}\right)} \Delta P_{\mathrm{m}} = \frac{\pi f}{H} \frac{1}{(s^2 + 2\xi s + \omega_{\mathrm{osc}}^2)} \Delta P_{\mathrm{m}} \qquad (8.8)$$

The natural oscillation frequency, ω_{osc}, is given by $\omega_{\mathrm{osc}} = \sqrt{(\pi f C_s/H)}$ and is therefore dependent on the inertia constant, H, frequency, f, and the synchronizing power coefficient, C_s.

The equivalent damping factor, ξ, is given by $\xi = C_d/4H$ and is also dependent on the inertia constant, H, and on the damping power coefficient, C_d.

Synchronous generators are designed for efficient operation and have low leakage and resistance losses and without the damping circuits, built into both the d axis and the q axis of the rotor, would possess little in the way of natural damping.

Currents flow in the damper circuits only under transient conditions. When the generator is operating under steady conditions, that is, with the rotor speed equal to the speed of rotation of the stator flux, the damper circuits do not cut the magnetic flux and therefore have zero current flow. However, when network oscillations or rotor speed oscillations are in evidence, the damper circuits of the rotor have relative motion with respect to the generator magnetic flux. As a consequence of the rotor speed being different from the stator flux speed, flux cutting occurs and in each damper circuit an emf is generated and current flows. Power is dissipated in the resistance of the circuit so that the energy of oscillation is reduced and damping of the oscillations is provided.

8.4 Influence of Automatic Voltage Regulator on Damping

The automatic voltage regulator (AVR) of a synchronous generator has two major functions. One is to maintain the generator terminal voltage close to its desired operating level as generator loading conditions change (DeMello and Concordia, 1969; Kundur, 1994). The other is to aid voltage recovery following a severe disturbance, such as a three-phase short-circuit on the network that is isolated via switchgear operation. While voltage control capability is very much dependent on the type of excitation system employed, all AVR excitation control schemes have to comply with Grid Code requirements in terms of basic performance.

Under oscillatory conditions, the currents in the generator damper windings are influenced by both the rotor speed oscillations and the changes in excitation voltage caused by the action of the AVR. The AVR therefore has an influence on damping and unfortunately it serves to reduce the natural damping of the generator. Although a full analysis of excitation control influence on damping is complex, the basic mechanism that causes the negative damping effect can be outlined as follows.

When the rotor angle increases, the generator stator current increases and the magnitude of the generator terminal voltage decreases due to the increased voltage drop across the generator reactance. The reduction in magnitude of the terminal voltage is sensed by the AVR, which causes the excitation system to increase the field voltage of the generator to produce an increase in the d axis flux and thereby an increase in the magnitude of the terminal voltage. Neither the excitation voltage nor the generator flux can be increased instantaneously, so that under oscillatory conditions the rotor flux variations lag the excitation voltage variations and hence the variations in rotor angle. In addition to its influence on terminal voltage, the flux increase also produces an increase in the generator torque and output power. Due to AVR action, therefore, a variation in generator power is produced that lags the oscillations in the rotor angle. The power variation under oscillatory conditions due to AVR control is shown vectorially in Figure 8.2. It can be seen that AVR action gives rise to a positive power component that is in phase with rotor angle oscillations, indicating a contribution that increases the synchronizing power. In addition, due to the lag between excitation voltage changes and flux changes, the AVR action also gives rise to a power component that is in anti-phase with rotor speed oscillations, indicating a negative contribution to generator damping.

The lag associated with the power variations can be reduced by the use of fast response excitation systems and by introducing a field current feedback signal into the excitation control loop that helps to reduce the effective time constant of the generator field. A suitably compensated, fast response excitation control scheme can help reduce the negative damping introduced by AVR control to a minimal level.

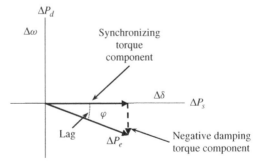

Figure 8.2   Vector diagram showing negative damping influence of an AVR

8.5   Influence on Damping of Generator Operating Conditions

Generator damping is dependent on the operating condition of the generator. The q axis damper circuits provide the main source of rotor damping and although the field and the d axis damper circuits do contribute to damping, their contribution tends to be significantly less.

The vector diagram in Figure 8.3 shows, in a simplified manner, the location of the generator flux with respect to the damper windings for a loaded operating condition. For very low values of leakage flux, the generator stator and rotor flux are approximately equal to one another and can be referred to simply as the generator flux, ψ. In cutting the stator windings, this rotating flux generates the stator voltage, E_t, and the per unit values of the magnitudes of flux and terminal voltage are effectively equal. When rotor speed oscillations occur, the q axis damper cuts the d axis component of the flux, ψ_d, and the d axis damper cuts the q axis component of the flux, ψ_q, so that emf is generated and current flows in the damper circuits.

The damping influence can be looked at from a power or a torque viewpoint. In the former, the current flow through the resistances of the damper circuits can be considered to dissipate the energy of the oscillation. In the latter, the interaction of the currents in the damper circuits with the respective flux components produces torques that are essentially in phase with rotor speed oscillations and consequently are damping torques.

In the vector diagram in Figure 8.3, the rotor angle, δ_r, is the angle between the d axis of the rotor and the flux, ψ, which is also the angle between the

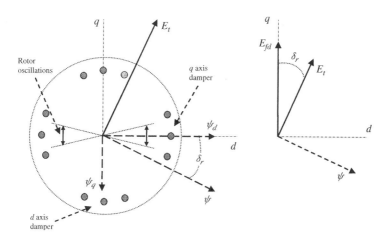

Figure 8.3   Flux cutting of generator damper windings

vectors E_{fd} and E_t. At low values of the rotor angle, δ_r, the generator flux, ψ, aligns closely with the d axis of the rotor, making the direct axis flux, ψ_d, very much greater than the quadrature axis flux, ψ_q. The emf generated in the q axis damper, due to rotor oscillations, is therefore very much greater than that generated in the d axis damper and this, coupled with the fact that the resistance of the q axis damper is much less than that of the d axis damper, leads to a much greater current flow in the q axis damper circuit and consequently a much greater value of damping torque (and damping power dissipation).

As the rotor angle, δ_r, increases, the d axis component of the flux, ψ_d, becomes smaller and the q axis component, ψ_q, becomes larger. Hence the emf generated in the q axis damper is reduced as the rotor angle is increased and the emf in the d axis is increased. However, due to the higher resistance of the d axis damper circuit, the increase in the d axis damper current does not compensate completely for the decrease in the q axis damper current and as a consequence the total damping torque is reduced. Generator damping therefore reduces as the rotor angle increases.

The operating rotor angle, δ_r, increases as the generator power output increases. For a round-rotor generator with synchronous reactances $X_d = X_q = X_s$, the rotor angle, δ_r, is given by

$$\delta_r = \sin^{-1} \frac{P_e X_s}{E_{fd} E_t} \tag{8.9}$$

Consequently, since an increase in power output is accompanied by an increase in the operating rotor angle, the damping torque provided by the rotor damper circuits decreases as the generator power increases. As can be seen from Figure 8.4, if the field voltage is kept constant, the rotor angle change is greater than in the case where high-gain AVR control is employed.

With constant field voltage, as the rotor angle increases the terminal voltage magnitude reduces, but with AVR control the terminal voltage magnitude is held approximately constant by increasing the magnitude of the field voltage. The lower rotor angle values with AVR control help to offset to a certain extent the negative damping introduced by the AVR control action.

The generator power factor also has an influence on the rotor angle value. For a fixed value of generator power output, the power factor is shifted from lagging to leading by reducing the generator field voltage. From Figure 8.5, this can be seen to result in an increase in rotor angle. Hence, for a given power output, the damping provided for lagging power factors is higher than that for leading power factors. With AVR control, the power factor is adjusted by changing the terminal voltage reference set point.

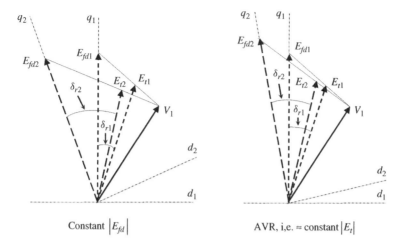

Figure 8.4   Change in rotor angle with increase in power output

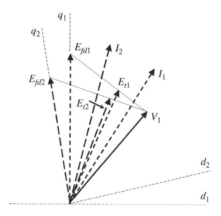

Figure 8.5   Vector diagram for leading and lagging power factors

8.6   Influence of Turbine Governor on Generator Operation

When the load on a power network increases, the increase in the load torques on the turbines leads to a reduction in the speed of the generators and hence a decrease in the network frequency. The governors of the turbines of generators allocated for frequency regulation duty sense the decrease in speed and increase the mechanical power outputs to enable the generators to supply the new level of the network load.

A schematic diagram showing the turbine–governor loop incorporated into the simplified synchronizing and damping torque model of the generator

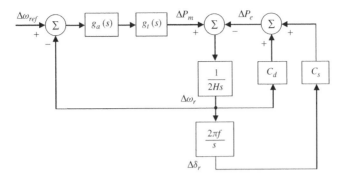

Figure 8.6   Governor–generator block diagram

is provided in Figure 8.6. When generator rotor oscillations occur, the action of the turbine governor, in response to the changes in speed, $\Delta\omega_r$, results in oscillatory variations in mechanical power, ΔP_m. Theoretically, if the governor–turbine system could provide an instantaneous response in mechanical power following speed changes, then under oscillatory conditions the turbine mechanical power oscillations would be in phase with the speed variations and, consequently, contribute directly to generator damping. In practice, the dynamic response of the governor loop of a steam turbine is relatively slow and the lags of the governor system and turbine produce a significant phase lag in response to rotor speed oscillations. Generally, for rotor oscillations in the local mode frequency region, the phase lag, φ, of the governor power loop is greater than $90°$.

In Figure 8.7, the phasor vector of mechanical power is split into two components, a synchronizing power component in phase with rotor angle oscillations and a damping power component in phase with rotor speed oscillations. When the phase lag, φ, is greater than $90°$, the damping power component can be seen to be negative, that is, in anti-phase with speed oscillations and therefore contributing negative damping.

At the lower frequencies corresponding to inter-area oscillations, the governor phase lag is lower and, if less than $90°$, the component of mechanical power in phase with speed will be positive and the governor will contribute positive damping.

Although a gas turbine has a much faster response than a steam turbine, the governor system of a gas turbine is purposely made slow to restrict demands on turbine output change, with the result that the combined phase lag of governor and turbine is similar to that of the steam turbine case. Consequently, the governor control of a gas turbine also introduces similar damping contributions under network oscillatory conditions.

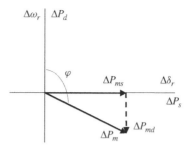

Figure 8.7   Vector diagram showing the negative damping contribution of the governor

In the case of a hydro turbine, the power response of the governor control loop is very slow. At network oscillation frequencies, the attenuation introduced by the slow response elements of the governor loop is so great that the turbine power output response can be considered negligible. A hydro turbine therefore has a negligible influence on network damping.

8.7   Transient Stability

A power network is a nonlinear system, so that although stability under small disturbance conditions is essential and well-damped response characteristics are highly desirable, these two items alone are not sufficient to ensure acceptable operation when large disturbances occur.

Following a large disturbance, such as a three-phase short-circuit on a transmission line that is quickly cleared by the action of the protection, the synchronous generators of the network must remain in synchronism with one another. The time taken for the protection system to isolate the faulted line via switchgear operation is specified by the Grid Code. For transmission system faults this is typically 80 ms, but for distribution networks the clearance time may be significantly longer. If the fault disturbance results in loss of synchronism, that is, pole slipping between generators, then the system is considered to be 'transiently unstable'. In terms of the definition of asymptotic stability, the system is not strictly unstable since after a period of pole slipping synchronous operation could be re-achieved. However, the huge current, voltage and power swings that accompany pole slipping do not constitute acceptable operation and could result in failure of the equipment if the offending generation were not tripped.

The fundamental characteristics of generator transient behaviour following a network fault can most easily be appreciated by considering a single machine feeding an infinite busbar load, as shown in Figure 8.8. The fault takes the

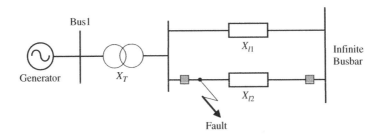

Figure 8.8   Single machine – infinite bus bar system

form of a three-phase short-circuit on the transformer end of one of the lines connecting the generator to the infinite busbar. The fault is isolated by opening the switchgear of the faulted line.

The power-angle characteristics of the pre- and post-fault system are shown in Figure 8.9. Starting from initial steady operating conditions (point A), consider the result of a three-phase short-circuit occurring on the line close to the transformer terminals.

During the 'fault-on' period, the generator power output (and therefore the load torque on the turbine) is reduced to zero. For simplicity, turbine governor action will be presumed to be sufficiently slow to permit the mechanical torque of the turbine to be considered constant. Hence, during the fault-on period,

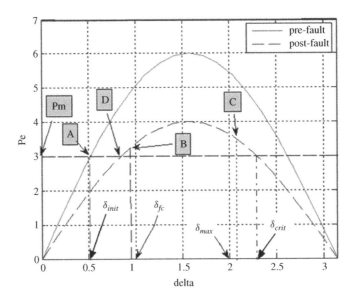

Figure 8.9   Pre- and post-fault generator power angle characteristics

the set accelerates and both the generator speed and load angle with respect to the infinite busbar increase in value.

Once the fault has been isolated, the generator can again feed power to the network and at fault clearance the operating point is that of B. Provided that the generator load torque is greater than the turbine driving torque, the set will start to decelerate. Throughout the period where the generator speed is greater than the synchronous value of the infinite busbar, the load angle will continue to increase. Provided that the load angle reaches a maximum value (point C) that is less than the critical value shown in Figure 8.9, the set will continue to decelerate and be driven back towards the new operating point D. If the load angle reaches the critical value and the speed is still higher than the synchronous value, then the set will continue to accelerate and synchronism will be lost. Pole slipping then occurs.

8.8 Voltage Stability

This generally is concerned with the ability of the system to accept additional load without voltage collapse. Generally, as load power increases, the current in the line feeding the load increases, creating an increase in the voltage drop across the line and a lower voltage at the load connection point.

In the network shown in Figure 8.10, where the line is purely inductive (of reactance jX) and the load is purely resistive, as additional resistive elements are added to the load the voltage–power characteristic has the form shown in Figure 8.11.

The maximum power that can be transmitted occurs when the resistance of the load R is equal to the reactance of the line. Beyond this, any further

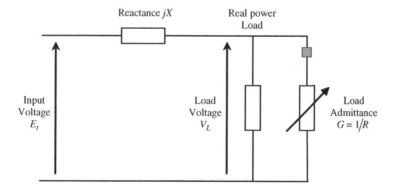

Figure 8.10 Resistive load fed via a purely reactive power line

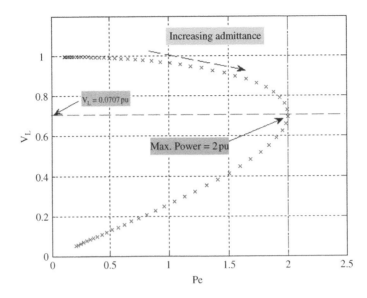

Figure 8.11 Voltage–power characteristic of a resistive load supplied via a reactive line

addition of resistive elements to the load results in a net reduction in the total load power supplied.

If, however, the load elements are resistive but have a constant power characteristic, any further attempt to increase the load beyond the maximum value defined by $P_L = V^2/2X$ will result in increased current demands that will cause voltage collapse. Operation of the system is normally maintained well away from such critical situations, but abnormal operating conditions can lead to the risk of voltage collapse.

Voltage problems are more common under transient conditions and tend to be due to excess reactive power demand. If, for example, a fault on the network leads to a sustained reduction in voltage levels, then the driving torques of the induction motors on the network will be reduced. This leads to a reduction in speed (i.e. an increase in slip) and an initial increase in the motor driving torque. If, however, the slip reaches the value corresponding to the maximum torque level available and this is less than the motor mechanical load torque, then the motor speed will collapse and the higher slip value will increase its reactive power demand and result in a further reduction in its terminal voltage. When large blocks of motors are involved, if the offending motors are not tripped then voltage collapse of the local network could result.

In the case of induction generators, as found on FSIG wind farms, if a fault on the network leads to a sustained reduction in voltage levels, the reduction in generator terminal voltage again leads to a reduction in generator torque levels and the power that can be transmitted to the network. When the wind turbine torque exceeds the load torque of the generator, the generator will accelerate, and if the speed corresponding to the maximum torque value is surpassed, then the set will suffer run-away and the increased reactive power demand, caused by the higher level of super-synchronous slip, will give rise to a further reduction in voltage levels and could again lead to voltage collapse on the local network if appropriate trip action is not carried out.

8.9 Generic Test Network

The generic network model used to demonstrate the influence of wind generation on network dynamic behaviour and transient performance is presented in Figure 8.12 (Anaya-Lara *et al.*, 2004). The generator and network data are chosen to be representative of a projected UK operating scenario with a large wind generation component sited on the northern Scotland network.

The model is implemented on Simulink, which enables dynamic stability to be assessed via eigenvalue analysis and transient performance to be assessed via the simulation of behaviour following a three-phase fault on the network.

Generator 3 is a steam turbine-driven synchronous generator provided with governor and excitation control. It is chosen to be representative of the main England–Wales network and has a rating of 21 000 MVA. Generator 1 is also a steam turbine-driven, synchronous generator provided with governor and excitation control. It is chosen to be representative of the southern Scotland network and has a rating of 2800 MVA. Generator 2 is chosen to represent a projected wind generation situation on the northern Scotland network and has

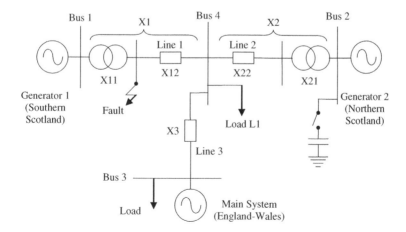

Figure 8.12 Generic network model

a rating of 2400 MVA. This generator can be a fixed speed induction generator (FSIG), a doubly fed induction generator (DFIG) or fully rated converter (FRC) employing an induction generator. Whereas generator 1 is closely coupled to the central busbar 4, generator 2 represents remote generation and the line to busbar 4 has a reactance of $X_{22} = 0.1337$ pu in comparison with line reactance $X_{12} = 0.01$ pu, both with respect to the system base of 1000 MVA.

When an FSIG is used, capacitive compensation is provided on the generator terminals in order to supply the reactive power demand of the FSIG while maintaining the desired voltage profile for the network. In the DFIG case, two distinct forms of control scheme are dealt with. The first controller, termed the PVdq scheme (current-based control scheme), controls terminal voltage via the manipulation of the d axis component of the DFIG rotor voltage and controls torque (or power) via the manipulation of the q axis component of rotor voltage. The second is called the flux magnitude and angle controller (FMAC) scheme. Here the magnitude of the rotor voltage is manipulated to control the terminal voltage magnitude and the phase angle of the rotor voltage is manipulated to control the power output. The latter provides lower interaction control than the PVdq scheme and lends itself more readily to the provision of network support, particularly with respect to voltage control and system damping. The FRC wind farm employs the control scheme introduced in Chapter 6.

In addition, the generation of the northern Scotland network can be provided by conventional synchronous generation having the same control provision as modelled for generators 1 and 3. This situation is used to provide a baseline case against which the influence of wind generation on the network dynamics can be evaluated. Basic generator and network data employed are provided in Appendix D.

8.10   Influence of Generation Type on Network Dynamic Stability

Eigenvalue results will now be presented with the object of showing how network dynamic performance is influenced as the wind generation component is increased (Anaya-Lara *et al.*, 2006). In the generic network in Figure 8.12, the installed generation capacity of generator 2 is increased in steps of 20% up to the maximum capacity considered of 2400 MVA. In addition to the wind generation situations, that where generator 2 is a synchronous generator is also included for comparative purposes.

8.10.1   Generator 2 – Synchronous Generator

The eigenvalue locus shown in Figure 8.13 for the case where generator 2 represents conventional synchronous generation assumes that all three synchronous generators have basic AVR control. When synchronous generator 2 has low capacity the system is stable, with the dominant network eigenvalue (corresponding to a local mode frequency) lying in the left-half plane.

As the capacity of generator 2 is increased, the frequency of the oscillatory mode is seen to decrease, indicating a reduction in the synchronizing power coefficients of generators 1 and 2. At lower frequencies, the damper circuits have lower generated voltages and lower currents and as a consequence their damping contribution decreases. At higher generation levels, the damping of the generator is insufficient to overcome the negative damping contributions of the governor and AVR and the oscillatory mode becomes unstable. Even for an installed capacity for generator 2 of 960 MVA, at a power level of 850 MW, the network is dynamically unstable.

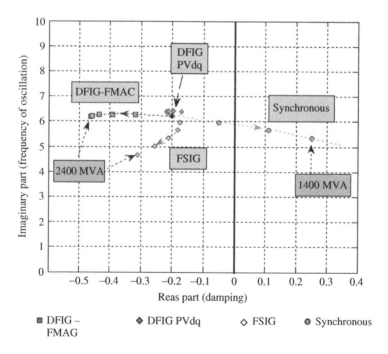

Figure 8.13   Influence of MVA rating of various type of generation (generator 2) on network damping (dominant eigenvalues)

8.10.2   Generator 2 – FSIG-based Wind Farm

When generator 2 is an FSIG-based wind farm, it can be seen from Figure 8.13 that at higher levels of wind generation dynamic stability actually improves. The response of the induction generator to system oscillations is to inject output current variations into the network that serve to increase current flow in the damper circuits of synchronous generator 1 and thereby provide increased damping.

8.10.3   Generator 2 – DFIG-based Wind Farm (PVdq Control)

With the DFIG-based wind generation employing the PVdq controller, the location of the dominant eigenvalue in the complex plane varies only slightly as the wind generation capacity is increased. As can be seen in the time responses of Figure 8.23, under network oscillatory conditions the power swings of the DFIG are approximately in anti-phase with those of synchronous generator 1. This reduces the phase angle variations of the voltage at the central busbar (bus 4) and this serves to reduce the variation in the synchronizing power coefficient of generator 1. As the mode frequency varies only slightly, the damping provided by generator 1 remains approximately constant.

8.10.4   Generator 2 – DFIG-based Wind Farm (FMAC Control)

When generator 2 is a DFIG with FMAC control, Figure 8.13 shows that the dominant eigenvalue is shifted progressively to the left in the complex plane as generation capacity is increased. Hence, unlike the PVdq scheme, with the FMAC scheme a positive contribution to the damping of synchronous generator 1 is made. This indicates that the control strategy adopted for a DFIG plays a significant role in the damping capability that can be provided.

8.10.5   Generator 2 – FRC-based Wind Farm

When generator 2 represents FRC-based wind generation, Figure 8.14 shows that as the capacity of the wind generation is increased, the mode frequency and the damping of the network are reduced. Unlike the cases with the other forms of wind generation, dynamic stability is lost well before the full capacity of 2400 MVA is reached.

Figure 8.14 indicates that with the FRC generation, network damping is slightly better than that for the synchronous generator case, where both generators 1 and 2 contribute negative damping at the higher generation capacity levels of generator 2. In general, by aiming for constant voltage and power output, the control of the FRC generation tends to render it dynamically

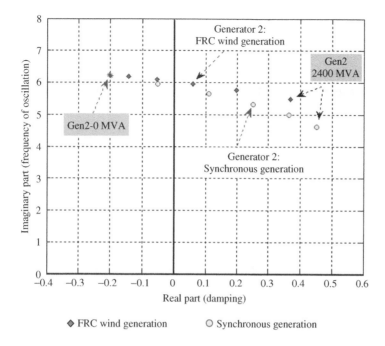

Figure 8.14 Influence of MVA rating of generator 2 on network dominant eigenvalues

neutral. As the capacity of the FRC generation is increased, the power transmitted to the main system increases, leading to an increased load angle for generator 1 and a reduction in its effective synchronizing power coefficient. This results in a reduction in the mode frequency and at lower oscillatory frequencies the voltages generated in the damper circuits are smaller, resulting in a reduction in their damping contribution.

8.11   Dynamic Interaction of Wind Farms with the Network

8.11.1   FSIG Influence on Network Damping

The concept of synchronizing torque and damping torque can also be used to reveal the interactive mechanism that enables a FSIG to provide a positive contribution to network damping.

The analysis to help explain the damping contribution will be based on the generic network of Figure 8.12. The busbar voltages of the network can be defined in terms of magnitude and their phase angle with respect to the system busbar, namely $E_1\angle\delta_1$, $E_2\angle\delta_2$, $E_3\angle\delta_3$ ($\delta_3 = 0$) and $E_4\angle\delta_4$. For simplicity, the magnitudes of the bus voltages will be considered to be constant so that the power variations through the lines are dependent solely on the phase changes of the voltages. It will also be assumed that under oscillatory conditions, the rate of change of the phase of the terminal voltage of generator 1 is directly dependent on the variation in its rotor speed. The turbine power output of each generator is taken to be constant.

Given the assumptions in the previous paragraph, the transmitted powers through lines 1, 2 and 3 are, respectively

$$P_{e1} = \frac{E_1 E_4}{X_1} \sin(\delta_1 - \delta_4) \tag{8.10}$$

$$P_{e2} = \frac{E_2 E_4}{X_2} \sin(\delta_2 - \delta_4) \tag{8.11}$$

$$P_{e3} = \frac{E_3 E_4}{X_3} \sin \delta_4 \tag{8.12}$$

For small variations, these can be converted to the form

$$\Delta P_{e1} = K_1(\Delta \delta_1 - \Delta \delta_4); \ \Delta P_{e2} = K_2(\Delta \delta_2 - \Delta \delta_4); \ \Delta P_{e3} = K_3 \Delta \delta_4 \tag{8.13}$$

where

$$K_1 = \frac{E_1 E_4}{X_1} \cos(\delta_{10} - \delta_{40}) \tag{8.14}$$

$$K_2 = \frac{E_2 E_4}{X_1} \cos(\delta_{20} - \delta_{40}) \tag{8.15}$$

$$K_3 = \frac{E_3 E_4}{X_3} \cos \delta_{40} \tag{8.16}$$

Since

$$\Delta P_{e3} = \Delta P_{e1} + \Delta P_{e2} \tag{8.17}$$

the relationship between the phases is given by

$$\Delta \delta_4 = \frac{K_1}{K_T} \Delta \delta_1 + \frac{K_2}{K_T} \Delta \delta_2 \tag{8.18}$$

where $K_T = K_1 + K_2 + K_3$.

In order to establish the basic interaction characteristics and identify how the dynamic behaviour of the FSIG influences the damping characteristics of synchronous generator 1, the result of a small increase in the rotor angle of generator 1 will be considered.

An increase in the rotor angle of generator 1 will produce an increase in phase angle, δ_1. The consequent increase in power flow through line 1, P_{e1}, will result in an increased power flow through line 3 that will increase the phase angle, δ_4, of the central busbar voltage with respect to the main system busbar. This increase in δ_4 will result in a reduction in the phase angle between the terminal voltage of generator 2 and the central busbar, $(\delta_2 - \delta_4)$ and will result in a reduction in the power flow in line 2 and hence a reduction in the power output of generator 2. Since the turbine power remains constant, this reduction in generator load power will result in the rotor accelerating, as dictated by the mechanics equation

$$\omega_{2r} = \frac{1}{2H_2 s}(P_{m2} - P_{e2}) \tag{8.19}$$

which for small deviations and constant mechanical power reduces to

$$\Delta\omega_{2r}(s) = \frac{1}{2H_2 s}(-\Delta P_{e2}) \tag{8.20}$$

This change in induction generator speed changes the slip value and results in a further change in the power output of generator 2.

The relationship between power output and slip can be approximated by $P_{e2} \approx -(E_2{}^2/R_r)s$. In per unit (pu) terms, slip $s = \omega_s - \omega_{2r}$, so that $\Delta s = \Delta\omega_s - \Delta\omega_{2r}$. For small variations, therefore,

$$\Delta P_{e2} \approx -\frac{E_2^2}{R_r}(\Delta\omega_s - \Delta\omega_{2r}) = K_s(\Delta\omega_s - \Delta\omega_{2r}) \tag{8.21}$$

It can be seen that slip is influenced by changes in the stator supply frequency. Since the rate of change of the phase, δ_2, of the terminal voltage represents a change in the stator frequency, this also has an influence on the slip and the power changes generated.

The stator frequency change (in pu) is given by

$$\Delta\omega_{2s} = \frac{s}{2\pi f}\Delta\delta_2 \tag{8.22}$$

A simplified model of the FSIG, which enables its power response, ΔP_{e2}, to a change in the phase of the central busbar, $\Delta\delta_4$, to be determined, is given in block diagram form in Figure 8.15.

The incorporation of this model within the network model relating power and angles gives rise to the block diagram in Figure 8.16. This enables the FSIG contribution to the dynamic power variations of generator 1, ΔP_{e1}, in

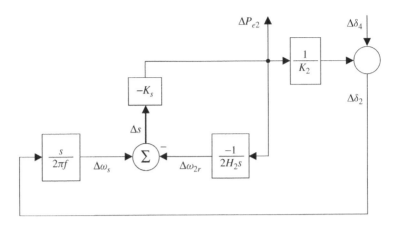

Figure 8.15   Simplified block diagram of an FSIG

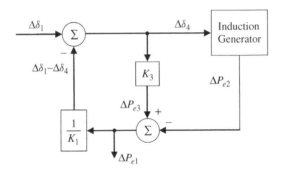

Figure 8.16   Simplified model for determining the interaction of FSIG generator 1 with the synchronous generator 2 of the generic network

response to oscillations in the phase of the terminal voltage of generator 1, $\Delta\delta_1$, to be determined.

For the assessment of damping contribution under oscillatory conditions, the behaviour of the model can be assessed in terms of its response at a specified oscillation frequency, ω_{osc}.

It should be noted that under oscillatory conditions, Eq. (8.20) can be re-expressed as

$$\Delta\omega_{2r} = \frac{j}{2H_2\omega_{osc}}\Delta P_{e2} \qquad (8.23)$$

This indicates that in vector form rotor speed, $\Delta\omega_{2r}$, leads generator power, ΔP_{e2}, by $90°$.

Also, for network oscillations at frequency ω_{osc}, the stator frequency changes are given by

$$\Delta\omega_{2s} = \frac{j\omega_{osc}}{2\pi f}\Delta\delta_2 \tag{8.24}$$

Making use of these relationships, at the specified frequency of concern, the dynamic model can be reduced to an algebraic set of equations. Solving the resulting equations for network oscillations of frequency $\omega_{osc} = 6$ rad s^{-1}, for the parameter values $K_1 = 16.66$, $K_2 = 5.235$, $K_3 = 5$, $K_s = 100$, $H = 3.5$ s and $f = 50$ Hz gives the following relationships for the power and phase angle variations:

$$\Delta P_{e1} = (3.4320 + j0.1806)\Delta\delta_1;$$
$$\Delta P_{e2} = (0.5383 - j0.2348)\Delta\delta_1; \quad \Delta\delta_4 = (0.7941 - j0.0108)\Delta\delta_1;$$
$$\Delta P_{e3} = (3.9704 - j0.0542)\Delta\delta_1 \quad \Delta\delta_2 = (0.8969 - j0.0557)\Delta\delta_1$$

The interactive relationships can be better appreciated by representing them pictorially in terms of the vector diagram in Figure 8.17. It should be noted that the vectors in the figure are not drawn to scale and are merely representative. In Figure 8.17, vector $\Delta\delta_1$ represents the oscillations in the phase angle of the terminal voltage of generator 1 and is drawn on the real axis.

Since the rate of change of the phase angle variations of the terminal voltage is directly dependent on the variation in the synchronous generator rotor speed, under oscillatory conditions the rotor speed vector, $\Delta\omega_1$, leads the phase angle vector, $\Delta\delta_1$, by 90° and therefore lies on the imaginary axis. The directions of the power vectors ΔP_{e1} and ΔP_{e2} are given from the angle

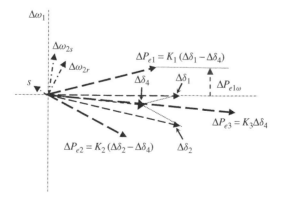

Figure 8.17   Vector diagram showing damping power influence of an FSIG

differences ($\Delta\delta_1 - \Delta\delta_4$) and ($\Delta\delta_2 - \Delta\delta_4$), respectively. The magnitude of the power vector ΔP_{e2} is dependent on the variations in slip and power vector ΔP_{e3} is given by the sum of the power vectors ΔP_{e1} and ΔP_{e2}. Power vector ΔP_{e1} can be seen to have a positive component, $\Delta P_{e1\omega}$, in the direction of the rotor speed vector, $\Delta\omega_1$. Hence, due to the presence of induction generator 2, synchronous generator 1 is provided with a component of electrical power that is in phase with its own rotor speed, indicating that the FSIG increases the damping of the synchronous generator when system oscillations occur.

8.11.2 DFIG Influence on Network Damping

The way in which a DFIG influences system damping can be again explained by making use of the concept of synchronizing torque and damping torque. The case where the DFIG of the generic network employs the FMAC control scheme will be used. In order to simplify the analysis, it will be assumed again that the magnitudes of the busbar voltages are constant so that power variations in the transmission lines are a function solely of phase angle variations.

Although the DFIG responds to variations in slip in a similar way to the FSIG, slip variation has a much smaller influence on the power variations than in the FSIG case. This is due to the fact that the constant, K_s, relating changes in output power to changes in slip is very much smaller in the case of the DFIG. A DFIG, with an operating value of slip of $s_0 = -0.1$, gives rise to a value of $K_s \approx 10$. In comparison, an FSIG having a typical operating slip value of $s_0 = -0.01$, has a value of $K_s \approx 100$. With a DFIG, the major influence on power output variations in response to network oscillations is the power loop of the controller.

In the DFIG, control over the voltage and power output is achieved by manipulation of the rotor voltage magnitude and its phase angle. Changes in V_r produce changes in the voltage behind transient reactance, E_g, generated in the stator and this influences the terminal voltage and power output of the DFIG. The response in E_g to changes in V_r is very rapid as can be deduced from the following DFIG equation [Eq. (5.12)]:

$$\frac{dE_g}{dt} = -\frac{1}{T_0}[E_g - j(X - X')I_s] + js\omega_s E_g - j\omega_s \frac{L_m}{L_{rr}}V_r \qquad (8.25)$$

The equations of the dynamic model of a DFIG are given in Chapter 4, which deals with FSIGs [Eqs (4.26) and (4.27)]. The above represents the vector form of the equation, rather than the individual d and q axis components.

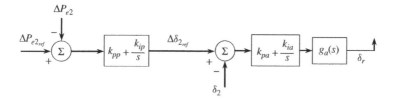

Figure 8.18 FMAC power control loop

In Eq. (8.25), $T_0 = 2.5$ s and since $\omega_s = 2\pi f$, the right-hand side of the equation is dominated by the last two terms. Very rapid response in $\mathbf{E}_g$ to changes in $\mathbf{V}_r$ is provided, so that for the purpose of damping analysis it can be assumed that $\mathbf{E}_g \approx (L_m/sL_{rr})\mathbf{V}_r$.

Consequently, in order to simplify analysis, it will be assumed that phase angle changes, δ_r, in the rotor voltage, $\mathbf{V}_r$, produce instantaneous changes in the phase angle, δ_g, of $\mathbf{E}_g$, the voltage behind transient reactance.

The power control loop of the FMAC scheme introduced in Chapter 5 in Figure 5.11 is shown in Figure 8.18.

The integral terms of the PI elements of the power loop have a relatively small influence on the controller output at network oscillatory frequencies and can be ignored for simplicity. Hence, for the damping analysis, the transfer functions of the power loop are approximated as $k_{pp} + (k_{ip}/s) \approx K_p; \ldots k_{pa} + (k_{ia}/s) \approx K$ and $g_a(s) = 1/(1 + sT)$. Further, as indicated earlier, the phase of the voltage behind transient reactance, δ_g, can be approximated by $\delta_g \approx \delta_r$, the phase of the rotor voltage.

The power output of the generator due to controller action can be calculated as

$$P_{ea} = \frac{E_g E_2}{X'} \sin(\delta_g - \delta_2) \tag{8.26}$$

where X' is the transient reactance of the DFIG, giving

$$\Delta P_{ea} = \frac{E_g E_2}{X'} \cos(\delta_{g0} - \delta_{20})(\Delta\delta_g - \Delta\delta_2) = K_g(\Delta\delta_g - \Delta\delta_2) \tag{8.27}$$

As in the FSIG case, the power change due to slip variation is given by

$$\Delta P_{eb} = K_s(\Delta\omega_s - \Delta\omega_r) \tag{8.28}$$

with $\Delta\omega_{2s} = (s/2\pi f)\Delta\delta_2$ and $\Delta\omega_{2r} = (1/2H_2s)\Delta P_{e2}$.

The power loop controller is simplified to

$$\Delta \delta_d = -k_p \Delta P_{e2}$$

$$\Delta \delta_r = \frac{K}{1 + sT}(\Delta \delta_d - \Delta \delta_r) \qquad (8.29)$$

$$\Delta \delta_g = \Delta \delta_r$$

The combined contribution to power change due control and slip influences is given by $\Delta P_{e2} = \Delta P_{ea} + \Delta P_{eb}$.

The simplified DFIG model with the FMAC scheme is shown in Figure 8.19.

This model can be incorporated as the induction generator model in the block diagram of Figure 8.16, to enable the DFIG contribution to the dynamic power variations of generator 1, ΔP_{e1}, due to variations in phase angle, $\Delta \delta_1$, to be determined.

For the assessment of the damping contribution under oscillatory conditions, the behaviour of the model can be assessed in terms of its response at the oscillation frequency of concern, ω_{osc}, where the dynamic model reduces to a set of algebraic relationships.

For a frequency of oscillation $\omega_{osc} = 6$ rad s^{-1} and with the parameter values $K_1 = 16.66$, $K_2 = 5.235$, $K_3 = 5$, $K_s = 10$, $H = 3.5$ s, $f = 50$ Hz, $K_p = 0.48$, $K = 6$, $T = 0.6666$ and $K_g = 12.66$. the solution of the model equations gives the following power and phase angle relationships

$$\Delta P_{e1} = (4.7031 + j0.2815)\Delta \delta_1; \qquad \Delta \delta_4 = (0.7178 - j0.0169)\Delta \delta_1;$$
$$\Delta P_{e2} = (-1.1140 - j0.3660)\Delta \delta_1; \qquad \Delta \delta_2 = (0.5050 - j0.0868)\Delta \delta_1$$
$$\Delta P_{e3} = (3.5891 - j0.0845)\Delta \delta_1$$

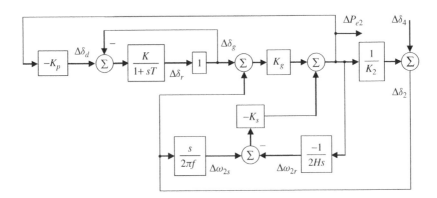

Figure 8.19   Simplified block diagram of a DFIG with FMAC control

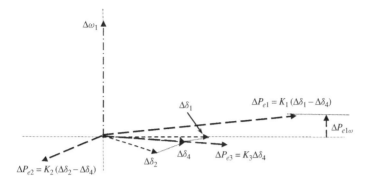

Figure 8.20 Vector diagram showing damping power influence of a DFIG

The positive imaginary component of ΔP_{e1} indicates that the DFIG contributes a component of power to synchronous generator 1 that is in phase with its rotor speed oscillations. Consequently, generator 1 is provided with an increase in its damping when network oscillations occur. The relationships between the powers and the phase angles are portrayed in the vector diagram in Figure 8.20. It should be noted that the vectors in the figure are not drawn to scale and are merely representative.

The major control contribution to damping is the lag term introduced into the power loop of the FMAC controller. Without the lag the ΔP_{e2} vector would align closely with the negative real axis and little damping contribution would be achieved.

Although the model employed to represent the network and generator is fairly crude, it does contain the major elements that influence damping and does help illuminate the way in which induction generators contribute to network damping.

8.12 Influence of Wind Generation on Network Transient Performance

A consideration of system behaviour following a three-phase short-circuit on the network close to the transformer of generator 1 is used to demonstrate the influence that the various types of generation have on network transient performance.

8.12.1 Generator 2 – Synchronous Generator

The responses shown as full lines in Figure 8.21 are for the case where only generators 1 and 3 are present on the network. It can be seen that generator 1

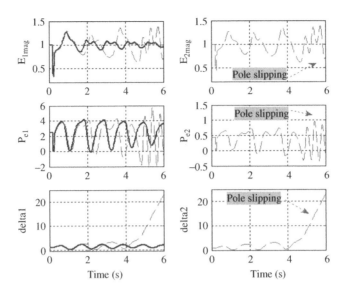

Figure 8.21 Fault near generator 1. Generator 2 is a synchronous generator of capacity 0 MVA (full lines) and 480 MVA (dashed lines)

remains in synchronism with the main system generator after fault clearance but the system damping is low.

When synchronous generator 2, of capacity 480 MVA, is introduced, following fault clearance synchronism is lost, with generators 1 and 2 remaining in synchronism with each other but losing synchronism with the main system generator 3. For the first few seconds after fault clearance the generator responses are oscillatory and, shortly after 4 s have elapsed, the rotor angles of both generators 1 and 2 increase continuously, indicating loss of synchronism and the resulting pole slipping gives rise to the higher frequency oscillations seen in the voltage and power responses.

The very poor post-fault behaviour is unacceptable and improved damping and transient performance need to be sought by the introduction of additional excitation control in the form of a power system stabilizer. This will be covered in the following chapter.

8.12.2 Generator 2 – FSIG Wind Farm

Although the eigenvalue analysis indicates that the network is dynamically stable for all generation capacities over the range 0–2400 MVA, the responses in Figure 8.22 show that, even for a generation capacity as low as 1440 MVA, fault ride-through is not achieved. In this case, the terminal voltage of the FSIG generator 2 fails to recover sufficiently following fault clearance, so that

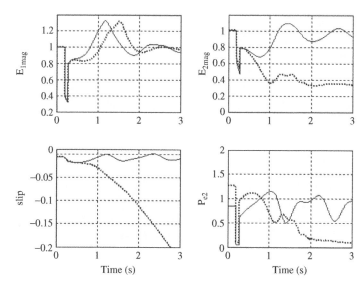

Figure 8.22   Fault near generator 1. Generator 2 is an FSIG wind farm of capacity 960 MVA (full lines) and 1440 MVA (dashed lines)

the maximum electrical torque level achievable is less than the wind turbine mechanical driving torque, and this leads to a further increase in the generator speed. As the slip becomes more negative, the reactive power demand of the generator increases and the increased current taken leads to a further reduction in the magnitude of the terminal voltage. The voltage collapses to less than 0.40 pu and the power transmitted to less than 10% of the initial value. These responses demonstrate a 'voltage instability' situation.

When the generating capacity is reduced to 960 MVA, following fault clearance the voltage recovery is sufficient to provide a generator load torque greater than that of the turbine mechanical driving torque. This enables the generator to decelerate and remain within the operating slip region and provide acceptable fault ride-through capability.

8.12.3   Generator 2 – DFIG Wind Farm

Figure 8.23 displays the post-fault performance when generator 2 is provided by DFIG wind farms. The performance when control is provided by the PVdq control scheme is compared with that when FMAC control is employed.

It can be seen that, for both types of control scheme, PVdq and FMAC, generators 1 and 3 remain in synchronism with one another following the fault, even with the maximum installed capacity of the DFIG wind generation (2400 MVA). This demonstrates the good fault ride-through capability

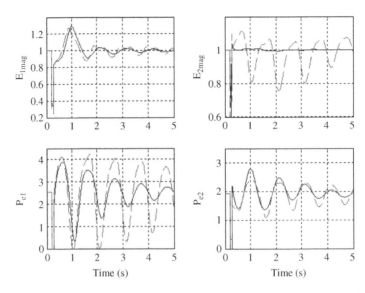

Figure 8.23 Fault near generator 1. Generator 2 is a DFIG wind farm with PVdq (dashed lines) and FMAC (full lines) control

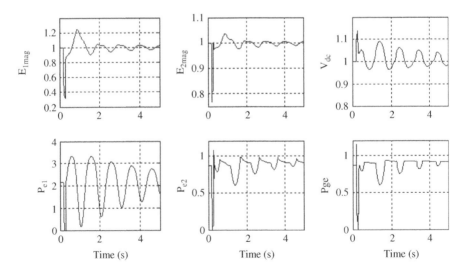

Figure 8.24 FRC wind generation – post-fault performance

of DFIG wind generation. The FMAC control case is seen to provide better post-fault damping, as was to be expected from the eigenvalue analysis.

8.12.4 Generator 2 – FRC Wind Farm

The eigenvalue analysis of the previous section showed that for an installed generation capacity of 1440 MVA and above, the network was dynamically unstable. The fault studies are therefore presented for the dynamically stable case of 960 MVA of installed FRC wind generation. The responses in Figure 8.24 show that the network synchronous generators retain synchronism following the fault, but the network damping is low. The swings in power and voltage indicate that additional system damping is required and the way that this can be achieved is covered in the following chapter.

References

Anaya-Lara, O., Hughes, F. M. and Jenkins, N. (2004) Generic network model for wind farm control scheme design and performance assessment, in *Proceedings of EWEC 2004 (European Wind Energy Conference)*, London.

Anaya-Lara, O., Hughes, F. M., Jenkins, N. and Strbac, G. (2006) Influence of wind farms on power system dynamic and transient stability, *Wind Engineering*, **30** (2), 107–127.

DeMello, F. P. and Concordia, C. (1969) Concepts of synchronous machine stability as effected by excitation control, *IEEE Transactions on Power Apparatus and Systems*, **PAS-88**, 316–329.

Grund, C. E., Paserba, J. J., Hauer, J. F. and Nilsson, S. (1993) Comparison of prony and eigenanalysis for power system control design, *IEEE Transactions on Power Systems*, **8** (3), 964–971.

IEEE/CIGRE Joint Task Force on Stability Terms and Definitions (2004) Definition and classification of power system stability, *IEEE Transactions on Power Systems*, **19** (2), 1387–1401.

Kundur, P. (1994) *Power System Stability and Control*, McGraw-Hill, New York, ISBN 0-07-035958-X.

Vittal, V. (2000) Consequence and impact of electric utility industry restructuring on transient stability and small-signal stability analysis, *Proceedings of the IEEE*, **88** (2), 196–207.

Wong, D. Y., Rogers, G. J., Porretta, B. and Kundur, P. (1988) Eigenvalue analysis of very large power systems, *IEEE Transactions on Power Systems*, **PWRS-3** (2), 472–480.

9

Power Systems Stabilizers and Network Damping Capability of Wind Farms

9.1 A Power System Stabilizer for a Synchronous Generator

9.1.1 Requirements and Function

The power system stabilizer (PSS) of a synchronous generator improves generator damping by manipulating its field voltage so that, in response to system oscillations, generator electrical power variations are produced that are in phase with rotor speed oscillations (Larsen and Swann, 1981; Kundur *et al.*, 1989).

The PSS can employ as its input signal any variable that responds to network oscillations. The output of the PSS, u_{pss}, is normally added to the reference set-point of the AVR excitation control loop as shown in Figure 9.1 (DeMello and Concordia, 1969; Kundur, 1994). The most commonly employed input signals are rotor speed and generator electrical power.

If, for the oscillation frequency of concern, ω_{osc}, the phase lag of the automatic voltage regulator, the excitation system and the generator, between the voltage reference set-point and electrical power output is θ_{eg}, then, for the case where the input signal of the PSS is generator rotor speed, ω, by designing the PSS such that, at frequency ω_{osc} it provides a phase lead of θ_{eg} between its input signal, ω, and its output, u_{pss}, the PSS will cause variations in electrical power to be generated that are in phase with rotor speed oscillations. The designed PSS control loop will then enable the excitation control to contribute directly to generator damping.

The generator response between field variations and power variations changes with the operating conditions and the network load. In addition,

Wind Energy Generation: Modelling and Control   Olimpo Anaya-Lara, Nick Jenkins, Janaka Ekanayake, Phill Cartwright and Mike Hughes
© 2009 John Wiley & Sons, Ltd

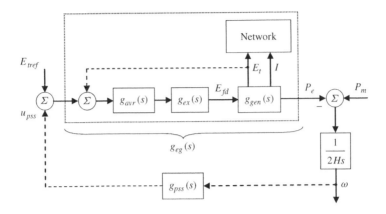

Figure 9.1   Generator excitation control with PSS

the frequency of the network oscillations also varies with the generator and network operating situation and a damping contribution needs to be provided over a frequency band that covers both local and inter-area oscillation frequencies. The phase compensation provided by the PSS needs to be designed so that a positive contribution to damping is provided across the foreseen bandwidth and range of operation. The phase lag of the combined transfer function of the AVR, excitation system and generator [$g_{eg}(s)$ in Figure 9.1], is considerably higher in the local mode frequency region than at the lower frequencies of inter-area mode oscillations. The phase lead compensator needs to be designed so that as the frequency of oscillation falls, the phase lead also falls to roughly match the fall in phase lag of $g_{eg}(s)$.

In terms of the vector diagram in Figure 9.2, it can readily be deduced that if the phase lead is lower than optimum, then, in addition to contributing to damping, the PSS will produce a negative contribution to synchronizing power. If the phase lead provided by the PSS is greater than the optimum value, then a positive contribution to synchronizing power is provided (Gibbard, 1988).

Turbine power variations due to governor action are normally sufficiently slow for the assumption of constant mechanical power to be made, so that under oscillatory conditions, the electrical power, P_e, lags speed by 90°, as shown in Figure 9.2. Hence, when electrical power is used as the input signal to the PSS, a negative gain is employed (effectively providing an input signal, $-P_e$, that leads speed by +90°) and phase lag compensation (of $\pi/2 - \theta_{eg}$) is required. This time, as the frequency of the oscillation of concern falls and the phase lag of $g_{eg}(s)$ falls, the compensator needs to be designed so that its phase lag increases and maintains the combined lag of the compensator and $g_{eg}(s)$ at approximately 90°.

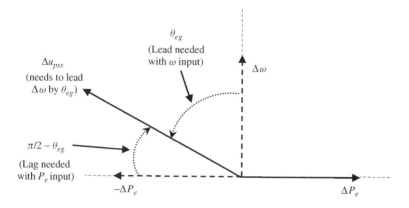

Figure 9.2   Vector diagram showing PSS compensator requirements

It should be pointed out that with synchronous generator excitation control, since both the AVR and the PSS exercise control by the manipulation of the same variable, namely the excitation voltage, independent control over both voltage and damping is not possible. Although a PSS can improve damping and extend the operating region of a synchronous generator, this is achieved at the expense of voltage control and leads to slower voltage recovery following network faults.

9.1.2   Synchronous Generator PSS and its Performance Contributions

9.1.2.1   Influence on Damping

The generic network introduced in the previous chapter will be used to demonstrate the influence that a PSS of a synchronous generator has on network dynamic performance and damping via eigenvalue analysis (Grund *et al.*, 1993). Figure 9.3 shows how the introduction of a PSS (based on an electrical power signal) into the excitation control of generator 1 improves network damping.

The generation capacity of generator 1 is 2800 MVA and operation is shown for the cases where the capacity of synchronous generator 2 is increased in steps of 20% from 0 to a maximum capacity of 2400 MVA (with the power range increasing from 0 to 2160 MW). Increasing the capacity of generator 2 has a considerable influence on network damping both with and without the PSS on generator 1. Although the PSS greatly improves damping at lower generating capacities of generator 2, for the full capacity of 2400 MVA the network is still seen to be unstable even with PSS control on generator 1.

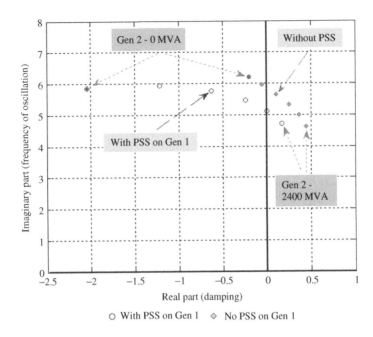

Figure 9.3 Influence of generator 1 PSS on network damping (dominant eigenvalues)

Figure 9.4 shows the loci of the dominant eigenvalues for the cases where PSS control is on (a) only generator 1, (b) only generator 2 and (c) on both generators 1 and 2. It can be seen that PSS control is needed on both generators 1 and 2 for dynamic stability to be preserved over the full capacity range of generator 2.

9.1.2.2   Influence on Transient Performance

The generic test network introduced in the previous chapter and shown in Figure 8.12 is again employed to demonstrate transient behaviour. A three-phase fault of duration 80 ms is applied on line 1 close to the transformer terminals of generator 1.

The eigenvalue plots in Figure 9.3 show that for the low-capacity situation of 480 MVA for generator 2, the network is stable without PSS control but has very low damping. The transient responses in Figure 9.5 show that, for this situation, following the fault both generators 1 and 2 lose synchronism with the main system and pole slipping occurs.

When a PSS is included in the excitation control of generator 1, following the fault the generators remain in synchronism and fault ride-through is achieved.

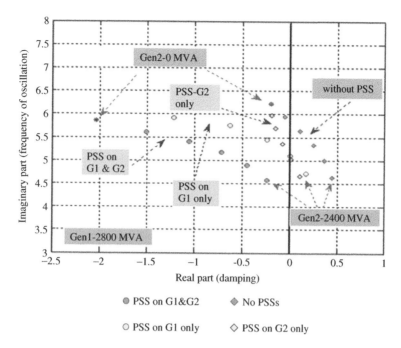

Figure 9.4 Influence of PSSs on generators 1 and 2 on network damping (dominant eigenvalues)

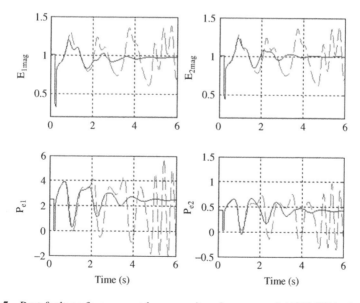

Figure 9.5 Post-fault performance at low capacity of generator 2 (480 MVA), showing the influence of a PSS on generator 1 (full lines) and without PSS (dashed lines)

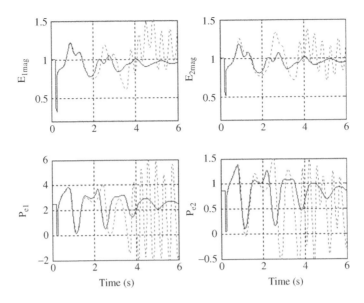

Figure 9.6   Post-fault performance with generator 2 capacity 960 MVA, showing the influence of PSSs on generators 1 and 2. PSS on generator 1 only (dashed lines); PSS on both
generators 1 and 2 (full lines)

When the capacity of generator 2 is increased to 960 MVA (Figure 9.6),
even with the PSS on generator 1, both generators lose synchronism with the
main system following the fault. The inclusion of PSSs on both generators 1
and 2 is necessary for fault ride-through to be achieved.

It can be seen that the inclusion of PSSs not only improves system damping but also helps to extend transient stability margins. However, when the
capacity of generator 2 is increased further to 1440 MVA, synchronism is
lost following fault clearance. The responses indicate that the generic test
network with only synchronous generation employed is very demanding on
synchronous generator excitation control.

9.2   A Power System Stabilizer for a DFIG

9.2.1   Requirements and Function

In the previous chapter, a simplified version of the generic test network
was used to show how power variations of FSIG- and DFIG-based wind
farms influence the behaviour and damping of the synchronous generation on
the local network. In the DFIG case, the damping contribution can be aug-
mented considerably by adding to the basic scheme an auxiliary PSS loop that,
under oscillatory network conditions, serves to inject power variations into the

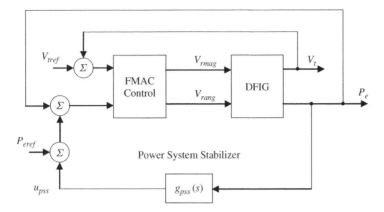

Figure 9.7   FMAC scheme of a DFIG with the PSS introduced at the power loop reference

network that stimulate additional damping power in the network synchronous generators (Hughes *et al.*, 2005, 2006).

The input signal to the PSS, theoretically, can be any local variable of the DFIG that responds to network oscillations, such as rotor speed, slip or stator electrical power. The output signal from the PSS, u_{pss}, can be introduced into the basic control scheme by adding it to the reference set-point of the power control loop as shown in Figure 9.7. As in the case of the PSS of a synchronous generator, the PSS of a DFIG consists of a wash-out term to eliminate steady-state offset and a phase shift compensator to provide the PSS output with the required phase relationship to improve damping.

The control requirement of a PSS for the DFIG of the generic network will now be assessed by making use of the concept of synchronizing power and damping power. The block diagram in Figure 9.8 shows the simplified model of a DFIG of the previous chapter when a PSS, based on a stator electrical power signal, is included.

In the simplified DFIG model with FMAC control introduced in Chapter 8, the power loop transfer function $g_p(s)$ can be expressed as

$$g_p(s) = \frac{K_p K'}{1 + T's} \tag{9.1}$$

where $K' = K/(1 + K)$ and $T' = T/(1 + K)$.

It is known that for the DFIG to contribute to network damping under oscillatory conditions, it is necessary for it to inject power oscillations into the network that engender power variations in the synchronous generators that are in phase with their rotor speed oscillations. In Chapter 8, it was shown that for

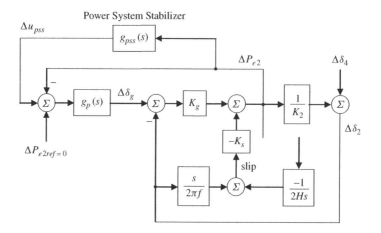

Figure 9.8   Simplified block diagram of a DFIG with a stator power-based PSS incorporated

this to take place the power variations of the DFIG must have a component of power that is in anti-phase with the rotor speed variations of the synchronous generation.

When the FMAC control is employed on the DFIG, the lag term included in the power loop ensures that a small but positive damping contribution is achieved. In terms of the generic network example, the vector diagram portrayed in Figure 9.9 indicates that if the PSS control serves to amplify the power vector, ΔP_{e2} and rotate it in an anticlockwise direction then the DFIG contribution to network damping will be increased. The component of the power vector, $\Delta P_{e2\omega}$, in the direction of the negative imaginary axis is increased and as a consequence the component of the power vector of generator 1 in the direction of the positive imaginary axis, $\Delta P_{e1\omega}$, also is increased. From the synchronous generator viewpoint, this represents an increase in the component of power in phase with its rotor speed oscillations, $\Delta\omega_1$, and hence an increase in damping power.

The following analysis is based on the simplified model of the DFIG and the generic network representation developed in the previous chapter.

In order to simplify the analysis, it will be assumed that the changes in power of the DFIG due to slip variations are sufficiently small to be ignored, that is, $K_s = 0$.

Then, from the simplified DFIG block diagram for the basic control scheme without the PSS, ΔP_{e2} is given by

$$\Delta P_{e2} = K_g[-g_p(s)\Delta P_{e2} - \Delta\delta_2] \tag{9.2}$$

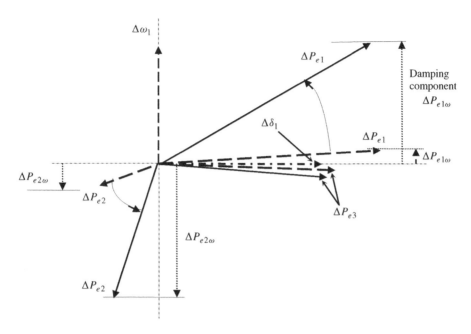

Figure 9.9   Vector diagram for generic network indicating the required influence of a PSS for improved damping of generator 1

and from the network relationships, ΔP_{e2} is also given by

$$\Delta P_{e2} = K_2[\Delta\delta_2 - \Delta\delta_4] \tag{9.3}$$

Hence, for the situation without the PSS, the relationship between generator power, ΔP_{e2}, and phase angle, $\Delta\delta_4$, can be expressed as

$$\left[\frac{1}{K_g} + \frac{1}{K_2} + g_p(s)\right]\Delta P_{e2} = -\Delta\delta_4 \tag{9.4}$$

If at oscillation frequency ω_{osc} the lag transfer function $g_p(s)$ can be expressed as

$$g_p(j\omega_{osc}) = A - jB \tag{9.5}$$

then letting

$$C = A + \frac{1}{K_g} + \frac{1}{K_2} \tag{9.6}$$

gives

$$\Delta P_{e2} = \frac{-1}{C - jB}\Delta\delta_4 = \frac{-C - jB}{C^2 + B^2}\Delta\delta_4 \qquad (9.7)$$

It has already been pointed out that to provide a contribution to damping, the power vector, ΔP_{e2}, needs to have a component that aligns with the speed variation, $-\Delta\omega_1$ (which is given by $-j\Delta\delta_1$). As phase vector $\Delta\delta_4$ aligns closely with phase vector $\Delta\delta_1$, this indicates that in Eq. (9.7) coefficient B should be positive and, with the FMAC scheme, since $g_p(s)$ is a lag transfer function, this is the case. Inspection of Eq. (9.7) indicates that in order to improve the damping of generator 1, the PSS should aim to increase coefficient B and reduce coefficient C. In terms of the vector diagram in Figure 9.9, this implies that the PSS should amplify vector ΔP_{e2} and rotate it in an anticlockwise direction.

When a PSS based on a stator power input signal ΔP_{e2} having the transfer function $g_{pss}(s)$ is introduced, the DFIG relationship of Eq. (9.4) becomes

$$\left\{\frac{1}{K_g} + \frac{1}{K_2} + g_p(s)[1 - g_{pss}(s)]\right\}\Delta P_{e2} = -\Delta\delta_4 \qquad (9.8)$$

If the transfer function $g_{pss}(s)$ is chosen to provide a phase lead characteristic such that, at the oscillation frequency of concern, the lead provided is greater than the lag of $g_p(s)$, then at frequency, ω_{osc}, the transfer function product $g_{pss}(s) \cdot g_p(s)$ can be expressed as

$$g_{pss}(j\omega_{osc}) \cdot g_p(j\omega_{osc}) = (A_{ps} + jB_{ps}) \qquad (9.9)$$

where both A_{ps} and B_{ps} are positive. Then Eq. (9.8) takes the form

$$\left[\frac{1}{K_g} + \frac{1}{K_2} + A - jB - (A_{ps} + jB_{ps})\right]\Delta P_{e2} = -\Delta\delta_4 \qquad (9.10)$$

Setting

$$C_p = \frac{1}{K_g} + \frac{1}{K_2} + A - A_{ps} \qquad (9.11)$$

and

$$B_p = B + B_{ps} \qquad (9.12)$$

the expression for power variation, ΔP_{e2} in terms of phase angle, $\Delta \delta_4$, becomes

$$\Delta P_{e2} = \frac{-1}{C_p - j B_p} \Delta \delta_4 = \frac{-C_p - j B_p}{C_p^2 + B_p^2} \Delta \delta_4 \qquad (9.13)$$

It can be seen that A_{ps} serves to reduce the coefficient C_p and B_{ps} serves to increase the coefficient B_p. Consequently, the negative imaginary component of ΔP_{e2} becomes greater in magnitude and since the phase angle $\Delta \delta_4$ aligns fairly closely with the phase angle $\Delta \delta_1$ of generator 1, this results in an increase in the component of power variation, ΔP_{e1}, of generator 1 that is in phase with its rotor speed oscillations.

In the analysis in the preceding section, a simplified representation of the generic network and DFIG was developed. The full generic network model was reduced to a level that included only the most influential basic dynamic mechanisms and interactions that influence network damping. The model was aimed at providing a simple basis for explaining the influence of DFIG and PSS on network dynamic behaviour and as such is not appropriate for detailed controller design purposes. Since the PSS influences dynamic modes not included in the simplified model, while forcing the eigenvalue pair associated with network mode oscillations significantly to the left in the complex plane, it also forces eigenvalues associated with other dynamic modes to the right. Consequently, a design based on the simplified model may well have an adverse and undesirable influence over dynamic modes ignored in the model. A more comprehensive system model needs to be employed in order to obtain a fuller indication of the overall influence of the PSS being designed. The scope of the simple model is therefore limited, but it can be used to establish basic design requirements.

As the DFIG improves network damping by enhancing the damping of the synchronous generation, its influence is indirect and an indirect design approach is favoured. Eigenvalue-based methods can readily accommodate comprehensive representations of the DFIG, its controller and the network and can be employed to design a PSS compensator that shifts the eigenvalues associated with network oscillations to desired locations while maintaining other system eigenvalues at acceptable locations in the complex plane.

A PSS based on stator power as an input signal, employing a phase lead compensator having the transfer function form and parameters given below was designed using an eigenvalue based approach.

$$g_{pss} = K_{ps} \frac{sT}{1 + sT} \frac{(1 + s T_b)^2}{(1 + s T_a)^2} \qquad (9.14)$$

where $K_{ps} = 1.0$, $T = 2.0$ s, $T_a = 0.05$ s and $T_b = 0.12$ s.

The way in which this PSS modifies the power and phase angle relationships of the simplified DFIG and generic network will now be considered.

Under oscillatory conditions, with a frequency of oscillation of $\omega_{osc} = 6\,\text{rad}\,\text{s}^{-1}$ and with the DFIG parameters $K_1 = 16.66$, $K_2 = 5.235$, $K_3 = 5$, $K_s = 10$, $H = 3.5\,\text{s}$, $f = 50\,\text{Hz}$, $K_p = 0.48$, $K = 6$, $T = 0.6666\,\text{s}$ and $K_g = 12.66$, the following power and phase angle relationships exist for the generic network when the basic FMAC control scheme is employed on the DFIG:

$$\Delta P_{e1} = (4.7031 + j0.2815)\Delta\delta_1;$$
$$\Delta P_{e2} = (-1.1140 - j0.3660)\Delta\delta_1; \quad \Delta\delta_4 = (0.7178 - j0.0169)\Delta\delta_1;$$
$$\Delta P_{e3} = (3.5891 - j0.0845)\Delta\delta_1 \quad \Delta\delta_2 = (0.5050 - j0.0868)\Delta\delta_1$$

When the designed PSS is incorporated, the relationships change to the following:

$$\Delta P_{e1} = (4.5574 + j1.5902)\Delta\delta_1;$$
$$\Delta P_{e2} = (-0.9247 - j2.0672)\Delta\delta_1; \quad \Delta\delta_4 = (0.7266 - j0.0954)\Delta\delta_1;$$
$$\Delta P_{e3} = (3.6327 - j0.4770)\Delta\delta_1 \quad \Delta\delta_2 = (0.5499 - j0.4903)\Delta\delta_1$$

A considerable increase is produced in the imaginary component of the $\Delta P_{e1}/\Delta\delta_1$ relationship, indicating that a significant contribution to network damping is achieved.

Also, when the calculated values of the network power and angle relationships are presented in vector diagram form as depicted in Figure 9.9, it can be seen that the desired vector manipulations have been achieved.

9.2.2   DFIG-PSS and its Performance Contributions

9.2.2.1   Influence on Damping

The generic network, for the case where generator 2 is a DFIG with FMAC control, will now be used to demonstrate the influence and capability of a PSS auxiliary control loop. The PSS is based on stator power as an input signal with the PSS output applied at the reference set point of the DFIG power control loop. The PSS compensator transfer function and parameters are those presented in the previous section.

The influence of increasing the PSS gain from zero to its design value is shown in Figure 9.10. The operating situation considered is that for 2400 MVA of DFIG generation operating at a nominal slip value of $s = -0.1$. It can be seen that as the gain is increased, the dominant eigenvalue, associated with network mode oscillations is shifted progressively to the left. Although a

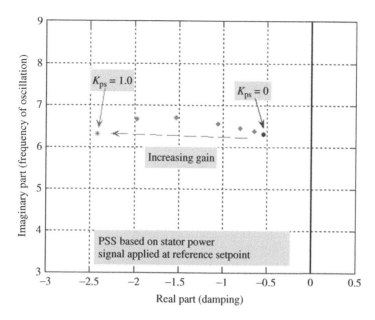

Figure 9.10 DFIG–FMAC control: influence of PSS gain on dominant eigenvalue. Capacity of generator 2 is 2400 MVA, $s_0 = -0.1$ pu

further increase in gain above $K_{ps} = 1.0$ would push this eigenvalue further to left of the complex plane, this design value was chosen as it provided good damping without compromising the location of other system modes over the operating range of the DFIG.

Figure 9.11 shows how the contribution to network damping is influenced as the DFIG generation capacity is increased. The DFIG generation has PSS control and operates at a slip value of $s = -0.1$ pu. As would be expected, the greater the installed capacity of DFIG generation, the greater is its relative damping influence on the synchronous generation of the local system.

Figure 9.12 shows how the operating slip value influences the damping power contribution of the DFIG PSS control. The case where the DFIG generation capacity is 2400 MVA is considered. As operation changes from super-synchronous to sub-synchronous slip values, that is, the operating speed of the DFIG decreases, the damping contribution is reduced. With a DFIG, at higher wind velocities and hence higher available wind power levels, higher operating speeds are employed to maximize the power transfer efficiency of the turbine. Consequently, at lower values of DFIG speed, the DFIG power output is lower and the influence of its manipulation by the PSS on synchronous generator damping is less.

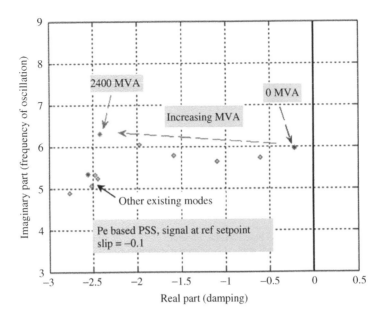

Figure 9.11   DFIG–FMAC control with PSS: influence of capacity of generator 2

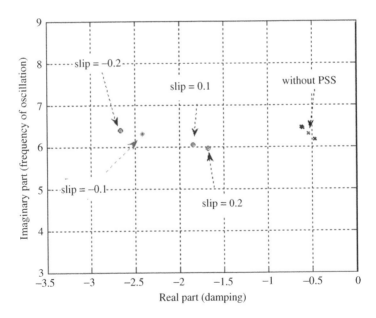

Figure 9.12   DFIG–FMAC control with PSS: influence of DFIG slip

9.2.2.2 Influence on Transient Performance

The generic test network will again be employed to demonstrate the influence of the PSS on transient performance. The situation where generator 2 represents 2400 MVA of DFIG generation with the availability of PSS control is considered. A three-phase fault of duration 80 ms is applied on line 1 close to the transformer terminals of generator 1.

In Figure 9.13, the dotted responses correspond to the case of the DFIG without the PSS, operating at a slip value of $s = -0.1$ pu. It can be seen that although fault ride-through is comfortably achieved, power oscillations persist beyond the 5 s period of the displayed response.

The full-line responses in Figure 9.13 correspond to the situation with PSS control included and these demonstrate that a considerable improvement is provided in the damping of the post-fault power transient of the synchronous generator. This damping improvement is achieved at the expense of increased power variation from the DFIG but with negligible change in the voltage recovery.

It should be pointed out that fault ride-through was not achieved for the synchronous generator case at the power level considered.

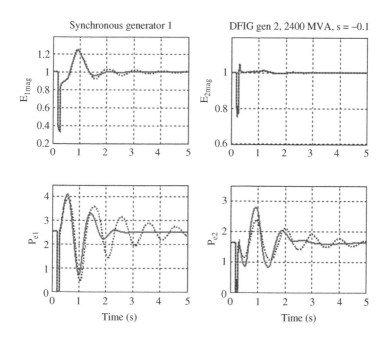

Figure 9.13 Influence of the PSS of a DFIG on post-fault performance. Super-synchronous operation with slip $s = -0.1$ pu. Dotted lines, without PSS; full lines, with PSS

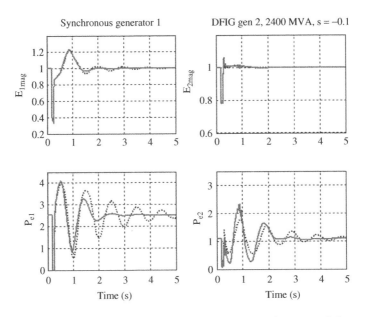

Figure 9.14 Influence of the PSS of a DFIG on post-fault performance. Sub-synchronous operation with slip $s = 0.1$ pu. Dotted lines, without PSS; full lines, with PSS

Figure 9.14 shows the equivalent case when a sub-synchronous operating slip of $s = 0.1$ pu is involved. The inclusion of the PSS again greatly improves the damping of the post-fault power transient of the synchronous generator. The damping of the power transient is in fact very similar to that for the $s = -0.1$ pu case, even though the operating power level is significantly lower. Over the initial portion of the post-fault period, signal levels are high and the limits of the PSS controller restrict the magnitude of its output. Consequently, over this period, the output of the PSS for the $s = 0.1$ pu case is almost identical with that for the $s = -0.1$ pu case, so that the variations in DFIG power output due to the PSS and the resulting influence over the synchronous generator response are also very similar.

9.3 A Power System Stabilizer for an FRC Wind Farm

9.3.1 Requirements and Functions

A fully-rated converter (FRC)-based wind farm employing tight, fast-acting control over voltage and power does not contribute to network damping under oscillatory conditions. Under normal operating conditions, the magnitude of the terminal voltage is kept essentially constant and although the power output will vary depending on the wind conditions, the aim of power control is to maintain wind farm operating conditions at desired levels and no attempt is made to provide network support by way of damping.

If an FRC wind farm is required to contribute to network damping, then additional control, in the form of a PSS, needs to be incorporated into the converter control scheme.

The generic network can again be employed to demonstrate the absence of a damping contribution when control is aimed solely at maintaining generator and converter conditions and the form that an auxiliary PSS loop should take if a contribution to network damping is to be provided.

Consider the FRC operating under constant turbine power conditions, with fast, tight control that maintains constant values of generator output voltage and power. As before, the voltages of all the network busbars will be considered to be constant, so that power variations in the lines are solely functions of the phase changes of the busbar voltages.

When rotor oscillations occur on synchronous generator 1, the power oscillations produced will cause oscillatory variations in the phase, $\Delta \delta_4$, of the voltage of busbar 4. The control of the FRC will maintain the output power of FRC generator 2 at a constant value, so that $\Delta P_{e2} = 0$. Since $\Delta P_{e2} = K_2(\Delta \delta_2 - \Delta \delta_4)$, this infers that $\Delta \delta_2 = \Delta \delta_4$.

The control action of the grid-side converter results in the phase of the terminal voltage of the FRC being adjusted to match the changes in the phase of busbar 4 and, with the difference between the phases of the voltages at either end of the line remaining constant, the power flow in line 2 remains constant.

In terms of the line relationships $\Delta P_{e1} = K_1(\Delta \delta_1 - \Delta \delta_4)$ and $\Delta P_{e3} = K_3 \Delta \delta_4$ and taking the idealized case where $\Delta P_{e2} = 0, \Delta P_{e3} = \Delta P_{e1} + \Delta P_{e2} = \Delta P_{e1}$ giving

$$\Delta \delta_4 = \frac{K_1}{K_1 + K_3} \Delta \delta_1 \qquad (9.15)$$

then with $K_1 = 16.66$, $K_2 = 5.235$ and $K_3 = 5$ this gives

$$\Delta \delta_4 = \Delta \delta_2 = 0.7692 \Delta \delta_1 \qquad (9.16)$$

so that $\Delta P_{e1} = 3.846 \Delta \delta_1 = \Delta P_{e3}$.

In terms of the vector diagram in Figure 9.15, vector ΔP_{e1} aligns with vector $\Delta \delta_1$, so that a component in the direction of the vector $\Delta \omega_1$ does not exist and therefore no contribution to damping is provided.

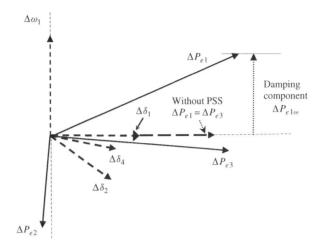

Figure 9.15   Vector diagram for the generic network showing how the PSS of an FRC can improve the damping of generator 1

In order to provide a contribution to damping, a suitable PSS input signal needs to be found that enables the FRC to respond to network oscillations. As the FRC control scheme aims to maintain constant output conditions, local variables such as voltage magnitude, current magnitude and power output have a negligible response to network oscillations, so that these variables exclude themselves as possible PSS input signals. A network signal is required that responds under oscillatory conditions and the variation in network frequency provides such a signal.

How a PSS based on a network frequency signal can enable an FRC generator to provide a network damping contribution can again be demonstrated using the simplified form of the generic network. In terms of the simplified generic network, a measure of network frequency is given by the rate of change of phase of the voltage of the central busbar, that is, $d(\Delta\delta_4)/dt$ (or in the Laplace domain, $s\Delta\delta_4$). If the PSS is considered to have the simple transfer function form $g_{pss}(s) = -K_{ps}$, then the power output variation of FRC generator 2 is given by

$$\Delta P_{e2} = -K_{ps}s\,\Delta\delta_4 \qquad (9.17)$$

and since $\Delta P_{e2} = K_2(\Delta\delta_2 - \Delta\delta_4)$, this gives

$$\Delta\delta_2 = \left(1 - \frac{K_{ps}}{K_2}s\right)\Delta\delta_4 \qquad (9.18)$$

For the network, $\Delta P_{e1} + \Delta P_{e2} = \Delta P_{e3}$, that is,

$$K_1(\Delta\delta_1 - \Delta\delta_4) + K_2(\Delta\delta_2 - \Delta\delta_4) = K_3\Delta\delta_4 \qquad (9.19)$$

giving

$$K_1\Delta\delta_1 + K_2\Delta\delta_2 = (K_1 + K_2 + K_3)\Delta\delta_4 = K_T\Delta\delta_4 \qquad (9.20)$$

so that

$$K_1\Delta\delta_1 = K_T\Delta\delta_4 - K_2\Delta\delta_2 = [(K_T - K_2) + sK_{ps}]\Delta\delta_4 \qquad (9.21)$$

For network oscillations of frequency ω_{osc}, $s = j\omega_{osc}$ giving

$$\Delta\delta_4 = \frac{K_1}{[(K_T - K_2) + j\omega_{osc}K_{ps}]}\Delta\delta_1 = (A - jB)\Delta\delta_1 \qquad (9.22)$$

where

$$A = \frac{K_1(K_T - K_2)}{C} \qquad (9.23)$$

$$B = \frac{K_1\omega_{osc}K_{ps}}{C} \qquad (9.24)$$

with

$$C = (K_T - K_2)^2 + (\omega_{osc}k_{ps})^2 \qquad (9.25)$$

The power variation of synchronous generator 1 is given by

$$\Delta P_{e1} = K_1(\Delta\delta_1 - \Delta\delta_4) = K_1(1 - A + jB)\Delta\delta_1 \qquad (9.26)$$

For a positive value of PSS gain K_{ps}, the coefficient B is positive and hence the coefficient K_1B of the imaginary term, jK_1B, of the power expression of Eq. (9.26) is also positive. Since $j\Delta\delta_1$ defines the direction of the rotor speed vector, $\Delta\omega_1$, of synchronous generator 1, a positive value of K_1B indicates that generator 1 has a power variation component that is in phase with its rotor speed oscillations. Hence, due to the introduction of the PSS into the control scheme of the FRC generation, the damping of the synchronous generator 1 is increased at the network mode frequency and the level of damping increases as the gain K_{ps} is increased.

For the case where $K_{ps} = 0.35$ and with $K_1 = 16.66$, $K_2 = 5.235$ and $K_3 = 5$, the following relationships exist for the simplified model:

$$\Delta P_{e1} = (4.5574 + j1.5902)\Delta\delta_1;$$
$$\Delta P_{e2} = (-0.9247 - j2.0672)\Delta\delta_1; \qquad \Delta\delta_4 = (0.7266 - j0.0954)\Delta\delta_1;$$
$$\Delta P_{e3} = (3.6327 - j0.4770)\Delta\delta_1 \qquad \Delta\delta_2 = (0.5499 - j0.4903)\Delta\delta_1$$

The relationships can again be depicted in vector diagram form, as shown in Figure 9.15, where the situations both with and without the PSS are portrayed.

9.3.2 FRC–PSS and its Performance Contributions

As the previous analysis indicated, a power system stabilizer based on a network frequency signal requires little in the way of phase compensation. A PSS was designed for the FRC of the generic network having the form

$$\Delta u_{pss} = -K_{ps}\frac{1 + sT_b}{1 + sT_a}\Delta f \qquad (9.27)$$

where $K_{ps} = 0.35$, $T_a = 0.085$ and $T_b = 0.75$, and the frequency variation Δf was obtained via a differentiator and filter:

$$\Delta f = \frac{s}{(1 + sT_{f1})(1 + sT_{f2})}\delta_4 \qquad (9.28)$$

where $T_{f1} = 0.05$ and $T_{f2} = 0.01$.

The PSS input signal, u_{pss}, is added to both the output power reference set point of the grid-side converter and the generator power reference set point of the generator-side converter. By manipulating the power demands of the grid side and generator-side converters in unison, the effect of the PSS demands on the converter DC voltage is minimized.

9.3.2.1 Influence on Damping

The way in which increasing the gain, K_{ps}, of the PSS influences the dominant (local mode) eigenvalue is shown in Figure 9.16. The operating situation considered is that where generator 1 represents 2800 MVA of synchronous generation and generator 2 represents 2400 MVA of FRC-based wind generation.

Without the PSS ($K_{ps} = 0$), the dominant eigenvalue lies in the right half of the complex plane, indicating that the network is dynamically unstable. As the gain, K_{ps}, is increased, the dominant eigenvalue is shifted progressively

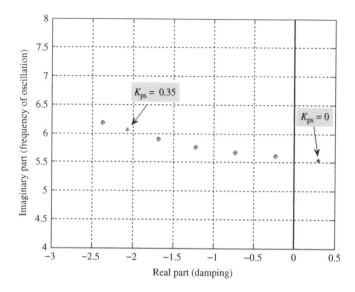

Figure 9.16 Influence of FRC–PSS gain, K_{ps}, on the dominant (local mode) eigenvalue

to the left, indicating that the damping torque of synchronous generator 1 is increased. The imaginary value, associated with the frequency of the oscillatory mode, increases with increase in gain, showing that an increase in the synchronizing torque of generator 1 is also produced. A further increase in the gain above the design value of 0.35 can push the eigenvalue significantly further to the left of the complex plane.

In designing the PSS, the gain value needs to be chosen to cater for the large disturbance conditions that occur following network faults, and also to provide a damping contribution. The greater the PSS gain, the greater is the impact of the PSS on the DC voltage level of the converters. The gain value of 0.35 was chosen so that for the fault disturbances considered for the generic network, the DC voltage of the converters did not change by more than ±10% from its nominal operating value. If greater damping and, consequently, a higher gain value for the PSS are required, a measure needs to be put in place to control the DC voltage variation within specified limits.

Figure 9.17 shows how the PSS damping provision is influenced by the FRC installed generation capacity and the operating power level. The PSS employed is that defined by Eqs (9.27) and (9.28). As the FRC generation capacity is increased, its capability to contribute to network damping via PSS control is increased. Figure 9.17 also shows that, at a given level of installed capacity, the FRC contribution to damping via PSS control increases as the operating power level of the generation increases.

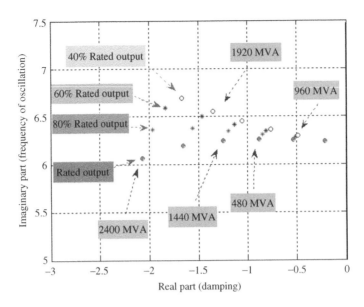

Figure 9.17   Influence of FRC generation capacity and operating power level on PSS damping contribution

9.3.2.2   Influence on Transient Performance

The generic test network is employed to demonstrate the influence that FRC generation with PSS control has on network transient performance. The result of the occurrence of a three-phase short-circuit on line 1, of duration 80 ms, close to the terminals of the transformer of generator 1 is simulated. The PSS introduced in the previous section is employed on the FRC.

Figure 9.18 shows the case where generator 1 represents 2800 MVA of synchronous generation having AVR control only and generator 2 represents 2400 MVA of FRC generation with PSS control. In addition to providing fault ride-through, very good damping is provided for the synchronous generator. This is in great contrast to the situation without PSS control on the FRC generation in the previous chapter, where fault ride-through was only achieved for an FRC generation capacity of 960 MVA.

Figure 9.19 provides a comparison of the contributions to network performance of the respective PSS control of FRC generation and synchronous generation. The case considered is that where generator 1 represents 2400 MVA of synchronous generation and generator 2 represents 1960 MVA of FRC generation. In Figure 9.19, the full-line responses correspond to the case where PSS control is on the FRC generation with the synchronous generation having only AVR control. The dotted-line responses correspond to the case where

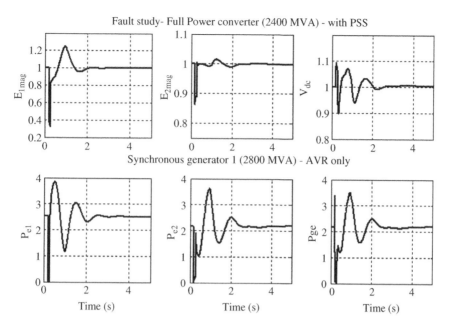

Figure 9.18 Network post-fault performance when FRC generation with PSS control is employed

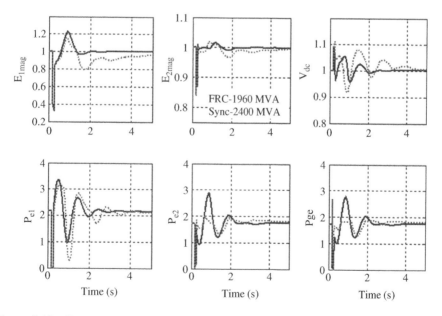

Figure 9.19 Comparison of PSS performance contributions. Full lines, PSS only on FRC generation; dotted lines, PSS only on synchronous generation

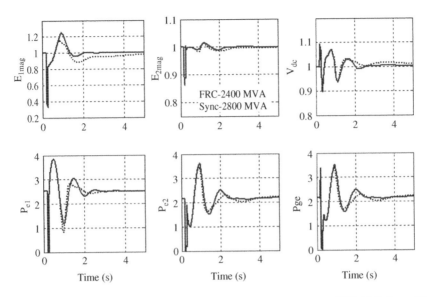

Figure 9.20 Compatibility of FRC–PSS with synchronous generator PSS. Full lines, FRC with PSS and synchronous generator with AVR only; dotted lines, FRC and synchronous generator with PSS

the FRC is without its PSS control and the synchronous generation has a PSS added to the basic AVR control. Compared with the case with PSS control only on synchronous generator 1, that with PSS control only on the FRC generation is seen to provide better performance for both generator 1 and generator 2 for the fault considered. Generator 1 is provided with better damping of power oscillations and shows better post-fault voltage recovery. In addition, the FRC generation, despite its PSS generating greater swings in output power, P_{2e}, has smaller deviations in both the terminal voltage magnitude, E_{2mag}, and the voltage of the DC link, V_{DC}.

A comparison could not be carried out for the capacity levels corresponding to Figure 9.18 since, with 2800 MVA of synchronous generation and 2400 MVA of FRC generation, fault ride-through was not achieved when PSS control was employed only on synchronous generator 1. A major reason for the poorer performance when a PSS is included only on the synchronous generator 1 is that, as generator 1 is closer to the fault, the signal level of its PSS is very high and controller limits consequently reduce its effectiveness and ability to contribute to generator performance.

The compatibility of the PSS control of FRC generation with the PSS control of synchronous generation is demonstrated by the responses in Figure 9.20. The case considered is again that with 2800 MVA of synchronous generation and 2400 MVA of FRC generation. The full-line responses are those corresponding to the situation where the FRC generation has PSS control and the dotted lines correspond to the situation where both synchronous and FRC generation have their respective PSSs in operation. Adding the PSS to the synchronous generation leads to a sharing of the damping power provision and reduced power swings are observed for both synchronous and FRC generation even though the damping level is increased.

References

DeMello, F. P. and Concordia, C. (1969) Concepts of synchronous machine stability as effected by excitation control, *IEEE Transactions on Power Apparatus and Systems*, **PAS-88**, 316–329.

Gibbard, M. J. (1988) Co-ordinated design of multimachine power system stabilizers based on damping torque concepts, *IEE Proceedings, Part C*, **135** (4), 276–284.

Grund, C. E., Paserba, J. J., Hauer, J. F. and Nilsson, S. (1993) Comparison of prony and eigenanalysis for power system control design, *IEEE Transactions on Power Systems*, **8** (3), 964–971.

Hughes, F. M., Anaya-Lara, O., Jenkins, N. and Strbac, G. (2005) Control of DFIG-based wind generation for power network support, *IEEE Transactions on Power Systems*, **20** (4), 1958–1966.

Hughes, F. M., Anaya-Lara, O., Jenkins, N. and Strbac, G. (2006) A power system stabilizer for DFIG-based wind generation, *IEEE Transactions on Power Systems*, **21** (2), 763–772.

Kundur, P. (1994) *Power System Stability and Control*, McGraw-Hill, New York, ISBN 0-07-035958-X.

Kundur, P., Klein, M., Rogers, G. J. and Zwyno, M. (1989) Applications of power system stabilizers for enhancement of overall system stability, *IEEE Transactions on Power Apparatus and Systems*, **4**, 614–622.

Larsen, E. V., Swann, D. A. (1981) Applying power system stabilizers, Part III, *IEEE Transactions on Power Apparatus and Systems*, **PAS-1000**, 3017–3046.

10

The Integration of Wind Farms into the Power System

In Europe, many future wind farms will be offshore, thus demanding power flow over highly capacitive cable networks. Further, the distance from the wind farm to the grid connection point is increasing. In the USA, India and China, transmission schemes from wind farms in excess of several hundred kilometres are being proposed. For transmitting bulk power over long distances or over capacitive networks, the developers and the system operators are facing a number of technical challenges. HVDC transmission and FACTS devices are being recognized as important possible enabling technologies.

10.1 Reactive Power Compensation

Early wind generators tended to use FSIGs, which consume reactive power and have limited controllability of real power (Holdsworth *et al.*, 2003). Because the Grid Connection Codes impose technical requirements, for example fault ride-through and also active and reactive power control capability, wind generator manufacturers have developed new systems, such as DFIGs and FRCs, which can meet these requirements. However, in order to satisfy the Grid Connection Code requirements at the grid connection point, wind farms may require the support of reactive power compensation devices such as SVCs and/or STATCOMs (Miller, 1982; Hingorani and Gyugyi, 2000; Acha *et al.*, 2001). The SVC/STATCOM can react to changes in the AC voltage within a few power frequency cycles and can thus eliminate the need for rapid switching of capacitor banks or transformer tap-changer operations. The rapid response of the SVC/STATCOM can also reduce the voltage drop experienced by the wind

Wind Energy Generation: Modelling and Control   Olimpo Anaya-Lara, Nick Jenkins,
Janaka Ekanayake, Phill Cartwright and Mike Hughes
© 2009 John Wiley & Sons, Ltd

farm during remote AC system faults, thus increasing the fault ride-through capability of the wind farm.

10.1.1 Static Var Compensator (SVC)

Reactive power compensation can be achieved with a variety of shunt devices. The simplest method is to connect a capacitor in parallel with the circuit. In many cases, it is connected to the output of the individual wind turbine generator via a circuit breaker or contactor and switched in and out when required. For the entire wind farm, a variable capacitance can be obtained by using a thyristor-controlled reactor (TCR) together with a fixed capacitor (FC) or by using a thyristor-switched capacitor (TSC). The TCR–FC gives smooth variation of capacitive reactive power support, whereas the TSC gives capacitive reactive power support in steps (Miller, 1982).

The basic elements of a TCR are a reactor in series with a bidirectional thyristor pair as shown in Figure 10.1a. The thyristors conduct on alternative half-cycles of the supply frequency. The current flow in the inductor, L, is controlled by adjusting the conduction interval of the back-to-back connected thyristors. This is achieved by delaying the closure of the thyristor switch by an angle α, which is referred to as the firing angle, in each half-cycle with respect to the voltage zero. When $\alpha = 90°$, the current is essentially reactive and sinusoidal. Partial conduction is obtained with firing angles between 90° and 180°. Outside the control range, when the thyristor is continuously conducting, the TCR behaves simply as a linear reactor. It is important to note that the TCR current always lags the voltage, so that reactive power can only be absorbed. However, the TCR compensator can be biased by a fixed capacitor, C, as shown in Figure 10.1b, so that its overall power factor can either be lagging or leading. The voltage–current characteristic of a TCR is shown in Figure 10.2.

In three-phase applications, the basic TCR elements are connected in delta through a transformer. The transformer is necessary for matching

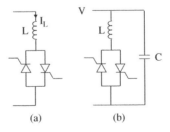

(a)   (b)

Figure 10.1 Thyristor-controlled reactor without and with a fixed capacitor

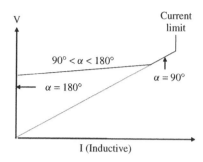

Figure 10.2   Voltage–current characteristic of a TCR scheme

the mains voltage to the thyristor valve voltage. This arrangement elim-
inates third-harmonic components produced by partial conduction of the
thyristors. Further elimination of harmonics can be achieved by using two
delta-connected TCRs of equal rating fed from two secondary windings of
the step-down transformer, one connected in star and the other in delta. This
forms a 12-pulse TCR. Moreover, the harmonics in the line current can be
reduced by replacing the fixed capacitors, associated with reactive power
generation, with a filter network. The filter can be designed to draw the
same fundamental current as the fixed capacitors at the system frequency and
provide low-impedance shunt paths at harmonic frequencies.

The basic elements of a TSC are a capacitor in series with a bidirectional
thyristor pair and a small reactor. The purpose of the reactor is to limit switching transients, to damp inrush currents and to form a filter for current harmonics
coming from the power system. In three-phase applications, the basic TSC
elements are connected in delta. The susceptance (1/reactance) is adjusted by
controlling the number of parallel capacitors connected in shunt. Each capacitor always conducts for an integral number of half cycles. The total susceptance
thus varies in a stepwise manner. The single-line diagram of the TSC scheme
is shown in Figure 10.3.

The output characteristic of a TSC is discontinuous and determined by the
rating and number of parallel-connected units. A smooth output characteristic
can be obtained by employing a number of TSC arms where capacitance is
set in binary manner ($C_1 = C, C_2 = 2C, C_3 = 4C, \ldots$) with TCR elements as
shown in Figure 10.4.

10.1.2   Static Synchronous Compensator (STATCOM)

A STATCOM is a voltage source converter (VSC)-based device, with the
voltage source behind a reactor (Figure 10.5). The voltage source is created

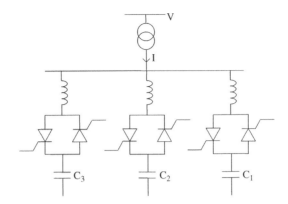

Figure 10.3   Single line diagram of a thyristor-switched capacitor (TSC)

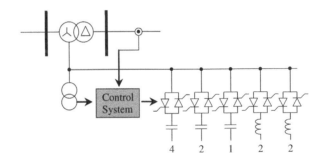

Figure 10.4   SVC using binary switched capacitors and switched reactors

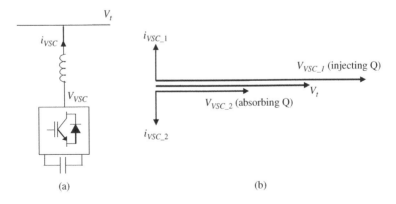

Figure 10.5   STATCOM arrangement. (a) STATCOM connection; (b) vector diagram

from a DC capacitor and therefore a STATCOM has very little real power capability. However, its real power capability can be increased if a suitable energy storage device is connected across the DC capacitor.

The reactive power at the terminals of the STATCOM depends on the amplitude of the voltage source. For example, if the terminal voltage of the VSC is higher than the AC voltage at the point of connection, the STATCOM generates reactive current; on the other hand, when the amplitude of the voltage source is lower than the AC voltage, it absorbs reactive power.

The response time of a STATCOM is shorter than that of an SVC, mainly due to the fast switching times provided by the IGBTs of the voltage source converter. The STATCOM also provides better reactive power support at low AC voltages than an SVC, since the reactive power from a STATCOM decreases linearly with the AC voltage (as the current can be maintained at the rated value even down to low AC voltage).

10.1.3 STATCOM and FSIG Stability

The STATCOM providing voltage control and a fault ride-through solution is demonstrated in this example by a simple model of a fixed-speed wind farm with STATCOM (Wu *et al.*, 2002). The case study and arrangement for the STATCOM are shown in Figure 10.6. The wind farm model consists of 30×2 MW FSIG, stall-regulated wind turbines. The short-circuit ratio (network short circuit level without wind farm connected/rating of wind farm) at the point of connection (PoC) of the wind farm is 10. For the simulation, it is assumed that the 132 kV network is subjected to a three-phase fault along one of the parallel circuits, of 150 ms duration at 2 s.

Without the STATCOM, the voltage at the point of connection does not recover to the pre-fault voltage after the clearance of the fault (Figure 10.7). However, when the STATCOM is set in operation, the wind farm is able to

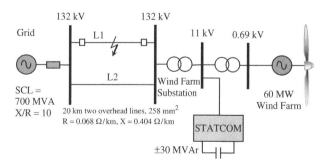

Figure 10.6 A large FSIG-based wind farm connected to the system with a STATCOM

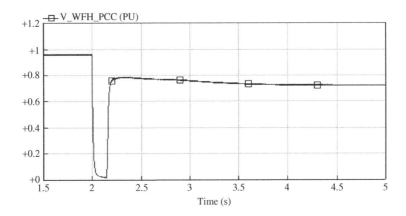

Figure 10.7   System voltage without STATCOM (Cartwright *et al.*, 2004)

ride through the fault as shown by the responses in Figure 10.8. During the
fault, the reactive power supplied by the STATCOM is decreased due to the
voltage drop (Figure 10.8b). After the fault, the STATCOM supplies reactive
power to the wind farm and compensates its requirements for reactive power
in order to ride through the fault.

10.2   HVAC Connections

HVAC will be used for most offshore wind farm applications with a connection
distance of less than ~50 km, where it will provide the simplest and most
economic connection method. Beyond a certain distance, the capacitive cable
current will approach the current rating of the cable and transmission of power
by means of AC is no longer economic (Sobrink *et al.*, 2007).

   At a connection distance in excess of 50 km, it is likely that reactive power
compensation will be required in order to keep the AC voltage amplitude
within the connection agreement requirements. At the transmission link end,
additional reactive power compensation may be required if the AC network is
relatively weak at this point, for example if the short-circuit level at the PoC
is less than three times the rating of the wind farm.

10.3   HVDC Connections

HVDC transmission may be the only feasible option for connection of an
offshore wind farm when the power levels are high and the cable distance is
long. HVDC offers the following advantages (particularly in those schemes
with cable connections) (Kimbark, 1971; Hammons *et al.*, 2000):

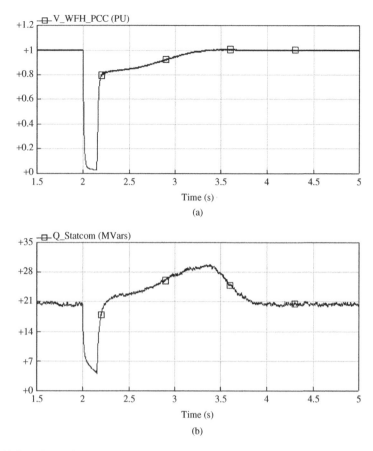

Figure 10.8 STATCOM in operation: (a) voltage (RMS) at PoC; (b) reactive power supplied by the STATCOM (Cartwright *et al.*, 2004)

- Wind farm and receiving grid networks are decoupled by the asynchronous connection, such that faults are not transferred between the two networks.
- DC transmission is not affected by cable charging currents.
- A pair of DC cables (currently used in LCC–HVDC schemes) can carry up to 1200 MW.
- The cable power loss is lower than for an equivalent AC cable scheme.

There are two different HVDC transmission technologies: line-commutated converter HVDC (LCC–HVDC) using thyristors and voltage source converter HVDC (VSC–HVDC) using IGBTs (Wright *et al.*, 2003). Further, multi-terminal HVDC schemes are also under consideration for offshore wind farm integration (Lu and Ooi, 2003).

Table 10.1   Capital costs for different transmission systems

Transmission Distance (km)	Total System Cost, HVAC (£m)	Total System Cost, LCC (£m)	Total System Cost, VSC (£m)
50	276	318	222
100	530	440	334
150	784	563	446
200	1,037	685	557
250	1,538	808	669
300	2,433	930	781
350	2,835	1,053	893
400	3,638	1,175	1,005

Table 10.1 shows costs associated with three options, HVAC, LCC–HVDC and VSC–HVDC, for a 1 GW offshore wind farm connection to the shore. The total cost of the installation was calculated using a cost model for each component involved in the system (cables, transformers, converter station, etc.) (Lundberg, 2003; Lazaridis, 2005). Costs for additional equipment such as the offshore platform required and the electrical power collection system for the wind farm were not considered.

From Table 10.1, it can be seen that all of the transmission systems begin at approximately the same cost, but (due to the cost of the cables) the cost of HVAC soon increases much more quickly than that of either of the HVDC systems. VSC–HVDC has the lowest capital cost of all the systems. However, the above analysis does not include the offshore platform that would be required by all the transmission systems. This platform would particularly affect the cost of LCC–HVDC as this transmission technology requires a large area for the converter station, which would lead to large, heavy and costly offshore platform.

10.3.1   LCC–HVDC

LCC–HVDC (Figure 10.9) can be used at very high power levels, with submarine cables suitable for up to 1200 MW (higher if parallel cables are used) being available. The reliability of LCC–HVDC has been demonstrated in over more than 30 years of service experience and the technology has lower power losses than VSC transmission. However, it is necessary to provide a commutation voltage in order for the LCC–HVDC converter to work. This commutation voltage has traditionally been supplied by synchronous generators or compensators in the AC network.

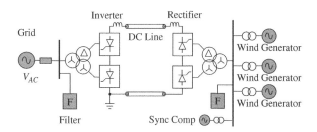

Figure 10.9   LCC–HVDC scheme with wind farm connection

With LCC – HVDC converters, current and power control is achieved by means of phase-angle firing control of the converters. To allow normal rectifier and inverter operation, as a rule the short-circuit level of the AC grid at the converter station should be at least 3 times the rated power of the DC link. Up to now the converter stations of an LCC – HVDC scheme have always been installed onshore.

In the event of a fault in the AC system close to the onshore terminal, power transmission will be interrupted until the fault has been cleared. After clearing the fault, the DC link may need 100–150 ms to return to full-power operation. However, the inertia of the wind generators can be used to store the wind energy during the transient interruption and the fault ride-through capability of the wind farm can be significantly improved. In order to permit fault recovery, the HVDC control scheme and that of the wind farm need to be closely coordinated.

10.3.2   VSC–HVDC

Many of the advantages described above for the LCC–HVDC scheme also apply to an VSC–HVDC scheme, in particular the asynchronous nature of the interconnection and the controllability of the power flow on the interconnector. The voltage source converter provides the following additional technical characteristics:

- It is self-commutating and does not require an external voltage source for its operation. Therefore, a synchronous generator or compensator is not needed at the offshore terminal in order to support the transmission of power.
- The reactive power flow can be independently controlled at each AC network. Therefore, AC harmonic filters and reactive power banks are not required to be varied as the load on the VSC transmission scheme changes.

- Reactive power control is independent of the active power control.
- There is no occurrence of commutation failure as a result of disturbances in the AC voltage.

These features make VSC–HVDC transmission attractive for the connection of offshore wind farms. Power may be transmitted to the wind farm at times of little or no wind and the AC voltage at either end can be controlled. However, VSC transmission does have higher power losses compared with an LCC–HVDC system. Further, its fault current contribution depends on the rating of the switches in the converters. For example, a thyristor (used in LCC–HVDC) can provide a short-term current (during the time of the fault) of 2–3 times its rated current, whereas the short-term current rating of an IGBT (used in VSC-HVDC) is more or less equal to its rating.

Figure 10.10 shows a typical arrangement of a wind farm connection using a point-to-point VSC–HVDC. The maximum power rating for a single converter at the end of 2005 was 330 MW, but up to 500 MW has been stated to be possible and several such systems may be used in parallel.

When an AC network fault occurs, the DC link voltages of the VSC–HVDC scheme will rise rapidly because the grid-side converters of the VSC–HVDC scheme are prevented from transmitting all the active power coming from the wind farm. Therefore, in order to maintain the DC link voltage below the upper limit, the excess power has to be dissipated or the power generated by the wind turbines has to be reduced. In order to dissipate the excess power, a chopper resistor is normally proposed (Akhmatov *et al.*, 2003; Ramtharan *et al.*, 2007). The method that can be used to reduce the power generated by the wind turbines depends on the turbine technology used. However, at the

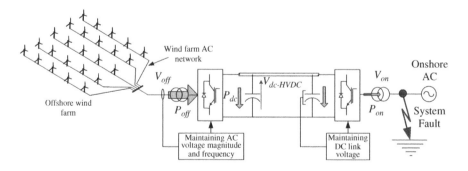

Figure 10.10   VSC–HVDC transmission scheme for wind farm application

turbine level the wind power has to be spilled using pitch control or has to be stored as kinetic energy in the rotating mass by increasing the rotor speed.

10.3.3 Multi-terminal HVDC

Multi-terminal HVDC technology has yet to be implemented widely, but it will permit the connection of multiple wind farms to a single DC circuit and/or multiple grid connections to the same DC circuit, as shown in Figure 10.11. Either LCC–HVDC, VSC–HVDC or hybrid technology could be employed in a multi-terminal configuration. In the case of LCC–HVDC-based multi-terminal systems, the converters are connected in series (Jovcic, 2008), whereas in the case of VSC–HVDC-based multi-terminal systems, the converters are connected in parallel (Figure 10.11) (Lu and Ooi, 2003).

There are basically two different control approaches that have been proposed for multi-terminal HVDC:

- *Master–slave*: One of the VSCs is responsible for controlling the DC bus voltage and the other VSCs follow a given power reference point which can be constant or assigned by the master converter. In the event of a failure on the main converter, another converter can take over the HVDC voltage control.
- *Coordinated control*: The VSCs control the DC bus voltage and power transfer in a coordinated manner. The HVDC voltage control can employ, for example, the droop-based technique where each converter has a given linear relationship between HVDC voltage and extracted power. Complex coordinated systems can be implemented with the use of communications,

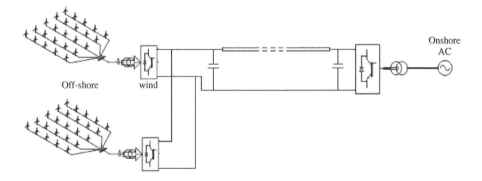

Figure 10.11 Multi-terminal wind farm based on HVDC–VSC technology

taking into account the different voltages and currents in the HVDC grid and seeking the optimum operating point for the overall system.

10.3.4 HVDC Transmission – Opportunities and Challenges

The technology required to embed HVDC schemes in an AC network is available today. However, there are a number of technical and economic challenges to be overcome in order to make HVDC transmission a viable option for wind farm connections and integration (Andersen, 2006).

10.3.4.1 Cost and Value of HVDC

The market for HVDC has traditionally been relatively small and there are very few manufacturers capable of providing such systems. With few projects realized, the benefits of mass production are not available and economic costs are not similar to those for conventional AC substation equipment of a similar rating.

10.3.4.2 Losses and Energy Unavailability

The power loss in an HVDC converter station is higher than that in an AC substation, because of the conversion from AC to DC and the harmonics produced by this process. However, the power loss in an HVDC transmission line can be 50–70% of that in an equivalent HVAC transmission line. Therefore, for large distances, an HVDC solution may have lower losses.

Energy unavailability is the amount of energy produced by the wind farm that cannot be transmitted to the onshore grid. This could be due to unexpected faults on the transmission system or to planned maintenance. By using figures for the likelihood of faults or maintenance on each component in a transmission system – typically averages of failures/required maintenance in installed components – a value for the energy unavailability can be calculated for each component (Lazaridis, 2005).

10.3.4.3 Transmission Cost

The transmission cost is the cost required to deliver a unit (kWh) of energy from the wind farm to the onshore grid. The energy delivered from the wind farm is calculated using the following equation:

$$E_D = E_P \times (1 - L) \times (1 - U) \qquad (10.1)$$

where E_D = energy delivered (MWh), E_P = energy produced (MWh), L = losses (%) and U = unavailability (%)

The analysis then assumed that a loan would be required to pay for the initial investment and from this the annual instalments for this loan could be calculated using

$$R = Investment \times \frac{r(1+r)^N}{(1+r)^N - 1} \tag{10.2}$$

where R = annual loan instalments (£m), $Investment$ = initial investment in the system (£m), r = loan interest rate (%), assumed to be 3%, and N = wind farm lifetime (years), assumed to be 30 years. From this, the cost of energy transmission can then be calculated from

$$Cost_{trans} = \frac{R}{E_D} \times \frac{1}{1-p} \tag{10.3}$$

where $Cost_{trans}$ = cost of energy transmission [£(kWh)$^{-1}$] and p = transmission system owner's profit (%), assumed to be 3%.

Figure 10.12 shows the transmission the cost for HVAC and HVDC up to 1000 km (Prentice, 2007).

These results show the unsuitability of HVAC as a means to transmit power from offshore wind farms that are located further than 50 km offshore. This is mainly due to losses which increase significantly with increased transmission

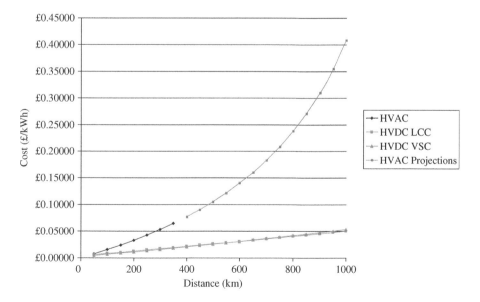

Figure 10.12   Transmission cost with the HVAC projections (top curve) included

distance. It is also easy to understand the transmission cost trends for both of the HVDC systems. VSC–HVDC is cheaper over shorter distances, but then, due to larger losses than LCC–HVDC, after 600 km LCC–HVDC becomes the most economic option.

10.3.4.4   Integration of an HVDC Scheme in an AC Network

Today, integration of an HVDC terminal into an AC system requires specialist engineering. The large AC harmonic filters, particularly for LCC–HVDC, can cause significant over-voltages during fault recovery, if the AC network strength is relatively weak. However, HVDC may result in improved performance during and after faults in the AC network and the performance can be optimized to suit particular network requirements through control system design.

The dynamic and transient performance of an HVDC scheme can be improved by the incorporation of dynamic reactive power capability. This capability is already available with VSC–HVDC transmission and could be added to LCC–HVDC, either through new circuit topologies and control algorithms or by the addition of new components, such as shunt or series reactive power compensation.

10.3.4.5   DC Side Faults

In a large point-to-point HVDC scheme, any fault on the DC side leads to shutting down of the connection until the fault is cleared and the system is restored. With multi-terminal HVDC networks, it is possible to take a section out of service during a converter or cable fault and leave the remaining system in operation. However, HVDC circuit breakers can be difficult to source and are likely to be expensive. There have been advances in HVDC circuit breaker technology in the recent past (Meyer *et al.*, 2005), but these have not yet become common industrial practice.

10.3.4.6   Stability of Network with Multi-terminal HVDC

If multiple HVDC schemes (which are point-to-point or multi-terminal) are to be used within a network, then the issue of interaction between these HVDC schemes would become increasingly important. Commutation failures, which are typically caused by voltage dips or sudden AC voltage phase-angle changes, could be caused by disturbances on another HVDC scheme and interaction between schemes could potentially cause instability, unless appropriate steps are taken. The problems are not insurmountable, as shown in several examples where HVDC converters terminate electrically close to each other and where good performance has been experienced. However, it is recommended that such systems are thoroughly modelled for specific scenarios.

It should be noted that VSC transmission does not suffer from commutation failures and is therefore not likely to suffer from instability, even if several HVDC schemes terminate in close proximity to each other.

10.4   Example of the Design of a Submarine Network

10.4.1   Beatrice Offshore Wind Farm

The proposed Beatrice offshore wind farm (OWF) is to be located 25 km off the coast of Scotland in the Moray Firth area (Figure 10.13). The wind turbines are to be placed approximately 5 km from the Beatrice Alpha oil platform in deep water (Beatrice wind, 2008).

There are currently two 5 MW prototype wind turbines installed in the area and it is proposed that this demonstration project be expanded to form the 1 GW Beatrice OWF. The connection to shore of this wind farm is very challenging as the onshore grid nearest to Beatrice is fairly weak and the circuits between Scotland and England (where most of the demand is) are already overloaded.

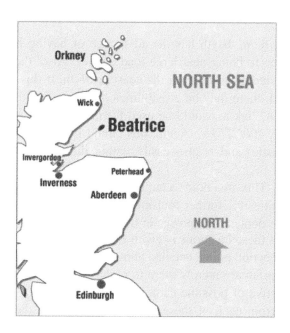

Figure 10.13   Location of the Beatrice offshore wind farm (Beatrice wind, 2008)

10.4.2 Onshore Grid Connection Points

To allow analysis of the onshore grid and to compare the effects of the different transmission types, onshore grid connection points had to be chosen (Prentice, 2007). Clearly, new substations could be constructed to allow for the connection of an OWF. However, a cost-effective method is to connect to existing substations and upgrade them if required. As the network in the North of Scotland is weak and the interconnectors between Scotland and England are already overloaded, an offshore transmission system was considered. It was proposed to connect the offshore network into points considerably further south of Beatrice. The offshore transmission network not only transfers power from Beatrice but also displaces some of the power that is currently transferred to England through the onshore network, thus relieving stresses on the latter.

10.4.2.1 Outline of Proposed Connection Points

Given the requirements above, six points on the East coast of the UK were selected in the study. The points selected are listed below along with a description and justification for selecting the connection point:

1. *Torness 400 kV substation area.* The Torness substation connects directly to Eccles, which is the Scottish end of one of the main Scotland–England interconnectors.
2. *Blyth 275 kV circuit.* Blyth has the advantage of having high-voltage circuits in addition to being near large load centres. Also, there are currently two offshore wind generators off the coast at Blyth. If this wind farm were to be expanded, there may be possibilities of connecting this wind farm to the same HVDC link as would be used for the Beatrice OWF.
3. *Hawthorne Pit. 400/275 kV.* Again, this circuit is very near to a large load centre (Newcastle) and is also easily within the vicinity of high-voltage circuitry.
4. *Grimsby West.* This area holds a large density of 400 kV circuits. There are good opportunities for further transmission to the south and there is also a relatively large demand for power in this area.
   A map showing these locations is given Figure 10.14. The thick black lines represent the connection points detailed above. It should be noted that the lines shown on the diagram are merely there to highlight the connection points and are not representative of possible cable routes.
   The distances from each of these connection points to the Beatrice OWF are given in Table 10.2. Each of these points was then analysed to ensure the suitability for the connection of additional generation in the area.

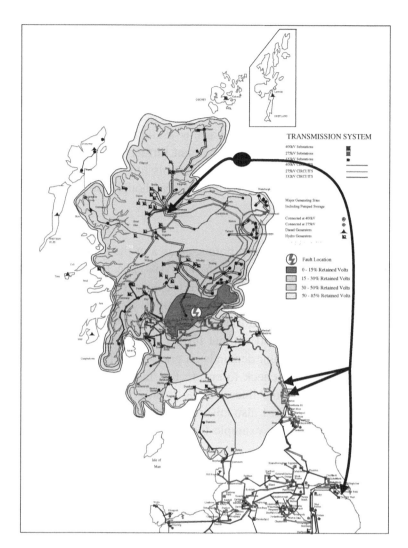

Figure 10.14 Location of points analysed for connection of the Beatrice offshore wind farm. The thick black lines represent the connection points detailed above

Table 10.2 Distances from the Beatrice site to substations

Substation	Distance to Beatrice (km)
Torness	300
Blyth	400
Hawthorne Pit	485
Grimsby West	670

The most suitable point for the connection of an offshore wind farm will be the optimal balance between the technical performance of the onshore grid and the financial and technical performance of the offshore transmission system. The onshore grid should be able to transmit the additional generation with little or no upgrades to either the transmission network or any plant involved (transformers, protection systems, etc.). A high demand within the vicinity of the connection point is also desirable as this would mean that the power could be used as soon as it comes onshore without further transmission/distribution being necessary. The capital cost of the transmission system should be as low as possible and the cost of energy transmission should also be low.

10.4.3 Technical Analysis

To analyse each of the connection points, an area of the grid around the actual connection substation was selected for analysis. The area was selected by determining a large substation which would represent an infinite grid connection point and had a large number of connections. It was assumed that if the local transmission system could transmit the power to these large substations, the power could be distributed without further strain on the grid. This technical analysis takes into account demands on grid supply points (GSPs) near the connection point and will check to ensure that there is a reasonable amount of additional capacity on the lines leading from connection point to a large substation. The analysis must also ensure that the $N-1$ security criterion, is met. Figure 10.15 shows the example for Torness.

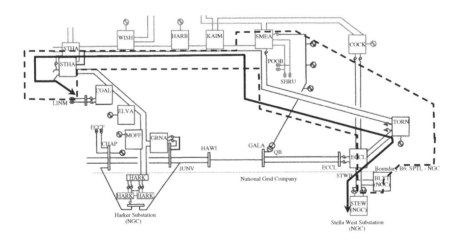

Figure 10.15   Grid area around Torness

10.4.3.1   Analysis of Transmission Lines

To analyse the additional capacity of the transmission lines, data for the ratings of each transmission line and the peak power flow along them were taken from the NGC 2006 Seven Year Statement (SYS) for the year 2006–2007 (National Grid, 2006). For the calculations, the transmission lines were assumed to be loaded to 75% of the circuit rating. Hence the maximum continuous loading was given as 75% of the maximum continuous circuit rating:

$$Loading_{max} = 0.75 \times Rating_{max} \quad \text{(MVA)} \quad\quad (10.4)$$

The peak winter power flow along each of the lines is given in MW in the SYS. To convert this value into MVA, a power factor (pf) of 0.97 was assumed:

$$Demand_{peak\,MVA} = \frac{Demand_{peak\,MW}}{pf} \quad\quad (10.5)$$

The additional capacity of each line could then be calculated by taking the maximum continuous load and subtracting the peak winter power flow. This then gives an additional power rating (in MVA) that could be sent along the transmission line:

$$Additional\ capacity = Loading_{max} - Demand_{peak\,MVA} \quad\quad (10.6)$$

From this analysis, it can be determined whether the 1000 MW of power from Beatrice can be transferred from the connection point or if reinforcements are required to allow this to be done. Table 10.3 shows the example for Torness.

10.4.3.2   Grid Supply Point (GSP) Demand

Values for the GSP demand were provided in the SYS. These values represented the demand on the transmission system from the connected distribution system at a certain substation. It was assumed that the power from Beatrice would be used to supply any demand at the grid supply points. Values are given in MW in the SYS and again the power factor was assumed to be 0.97. Table 10.4 gives the GSP demand in the Torness case.

10.4.3.3   The $N - 1$ Criterion

In this section, each of the areas are inspected to ensure they are capable of transmitting the supplied power. Table 10.5 shows the area being considered and details what piece of equipment has been removed to allow the analysis of the $N - 1$ criterion.

Table 10.3   Distances from Beatrice site to substations (Torness case)

From	To	$Rating_{max}$ (MVA)	$Loading_{max}$ (MVA)	$Demand_{peak}$ (MW)	$Demand_{peak}$ (MVA)	Additional capacity on-line (MVA)
TORN	SMEA	1250	937.5	−134	138.1	799.4
	SMEA	1130	847.5	164	169.1	678.4
	ECCL	1250	937.5	484	499.0	438.5
	ECCL	1250	937.5	484	499.0	438.5
SMEA	STHA	1390	1042.5	164	169.1	873.4
ECCL	STWB	1390	1042.5	660	680.4	362.1
	STWB	1390	1042.5	655	675.3	367.2
STHA	HAKB	2010	1507.5	806	830.9	676.6
	WISH	762	571.5	107	110.3	461.2
	LINM	1380	1035.0	410	422.7	612.3
KILS	STHA	1390	1042.5	92.1	94.9	947.6
INKI	STHA	1390	1042.5	316	325.8	716.7

Table 10.4   GSP demand (Torness case)

Substation name	Demand at GSP(MW)	Demand at GSP (MVA)
Torness	0	–
Eccles	28.4	29.3
Smeaton	0	–
Strathaven	54	55.7

Table 10.5   Analysis of $N-1$ criterion

Substation	Equipment removed	Result
Torness	TORN–ECCL double circuit	Failure. Insufficient capacity to transfer power
Blyth	BLYT–TYNE double circuit	Failure. Insufficient capacity to transfer power
Hawthorne Pit	HAWP–NORT circuit	Sufficient capacity to withstand connection
Grimsby West	KEAD–CREB double circuit	Sufficient capacity to withstand connection

10.4.4   Cost Analysis

The total cost of installation of the transmission system, by calculating the costs of the individual components and then calculating the total cost of the transmission system was determined.

10.4.5 Recommended Point of Connection

From the analyses above, it was possible to determine the optimum point of connection for the Beatrice OWF following the methodology shown in Figure 10.16. This methodology was used to produce the results shown in Table 10.6.

From these results, it is clear that the optimum connection point (determined in the study from the points analysed) was *Hawthorne Pit*. This connection point is the nearest connection point to Beatrice that is technically suitable to cope with the connection of new generation. This means that no upgrades would be required to the onshore grid and the cables would be as short as possible, leading to the lowest transmission system capital costs.

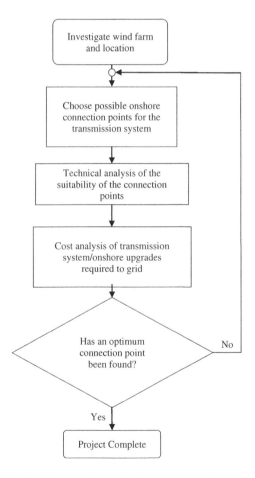

Figure 10.16 Methodology to determine the recommended point of connection

Table 10.6 Results of the study of the connection points

Substation	Additional capacity	GSP demand	Capital cost (£m)			Transmission cost [£ (kW h)$^{-1}$]		
			HVAC	HVDC LCC	HVDC VSC	HVAC	LCC HVDC	VSC HVDC
Torness	Deficit	Low	2427.98	930.29	781.08	0.0531	0.0169	0.0157
Blyth	Low	Medium	3231.58	1175.26	1004.77	0.0769	0.0214	0.0205
Hawthorne Pit	Medium	High	3914.64	1383.48	1194.90	0.1008	0.0253	0.0248
Grimsby West	High	High	5401.30	1836.67	1608.72	0.1691	0.0339	0.0345

Acknowledgement

The authors would like to acknowledge Mr. Colin Prentice for the support provided in the preparation of Section 10.4.

References

Acha, E., Agelidis, V. G., Anaya-Lara, O. and Miller, T. J. E. (2001) *Electronic Control in Electrical Power Systems*, Butterworth-Heinemann, London, ISBN 0-7506-5126-1.

Akhmatov, V., Nielsen, A., Pedersen, J. K. and Nymann, O. (2003) Variable-speed wind turbines with multi-pole synchronous permanent magnet generators. Part 1. Modelling in dynamic simulation tools, *Wind Engineering*, **27**, 531–548.

Andersen, B. R. (2006) HVDC transmission – opportunities and challenges, presented at the IEE International Conference on AC and DC Power Transmission, 28–31 March 2006.

Beatrice wind (2008) *The Beatrice Wind Farm Demonstrator Project*, available at http://www.beatricewind.co.uk/home/; last accessed 3 April 2009.

Cartwright, P., Anaya-Lara, O., Wu, X., Xu, L. and Jenkins, N. (2004) Grid compliant offshore wind power connections provided by FACTS and HVDC solutions, in *Proceedings of the European Wind Energy Conference EWEC, London*.

Hammons, T. J., Woodford, D., Loughtan, J., Chamia, M., Donahoe, J., Povh, D., Bisewski, B. and Long, W. (2000) Role of HVDC transmission in future energy development, *IEEE Power Engineering Review*, **20** (2), 10–25.

Hingorani, N. G. and Gyugyi, L. (2000) Understanding FACTS: Concept and Technology of Flexible A.C. Transmission Systems, Wiley-IEEE Press, New York, ISBN 0-7803-3455-8.

Holdsworth, L., Wu, X., Ekanayake, J. B. and Jenkins, N. (2003) Comparison of fixed speed and doubly-fed induction wind turbines during power system disturbances, *IEE Proceedings Generation, Transmission and Distribution*, **150** (3), 343–352.

Jovcic, D. (2008) Offshore wind farm with a series multi-terminal CSI HVDC, *Electric Power Systems Research*, **78**, 747–755.

Kimbark, E. W. (1971) *Direct Current Transmission*, Vol. 1, John Wiley & Sons, Inc., New York.

Lazaridis, L. P. (2005) Economic comparison of HVAC and HVDC solutions for large offshore wind farms under special consideration of reliability. Masters Thesis. Royal Institute of Technology, Stockholm.

Lu, W. and Ooi, B. T. (2003) Optimal acquisition and aggregation of offshore wind power by multi-terminal voltage-source HVDC, *IEEE Transactions on Power Delivery*, **18** (1), 201–206.

Lundberg, S. (2003) *Performance Comparison of Wind Park Configurations*, Technical Report, Department of Electric Power Engineering, Chalmers University of Technology, Gothenburg.

Meyer, C., Kowal, M. and De Doncker, R. W. (2005) Circuit breaker concepts for future high-power dc applications, in *Fourteenth IAS Annual Meeting Industry Applications Conference*, Vol. 2, pp. 860–866.

Miller, T. J. E. (1982) *Reactive Power Control in Electric Systems*, John Wiley & Sons, Inc., New York.

National Grid (2006) *National Grid Seven Year Statement*, National Grid Company (NGC) Ltd, available at http://www.nationalgrid.com/uk/electricity/; last accessed 3 April 2009.

Prentice, C. (2007) *Designing a submarine network for connection of offshore wind power plants, Internal Report*, Department of Electronics and Electrical Engineering, University of Strathclyde.

Ramtharan, G., Anaya-Lara, O and Jenkins, N. (2007) Modelling and control of synchronous generators for wide-range variable-speed wind turbines, *Wind Energy*, **10** (3), 231–246.

Sobrink, K. H., Woodford, D., Belhomme, R. and Joncquel, E. (2003) A.C. cable v D.C. cable transmission for offshore wind farms, a study case, presented at the 4th International Workshop on Large-scale Integration of Wind Power and Transmission Networks for Offshore Wind Farms, Billund, Denmark.

Wright, S. D., Rogers, A. L., Manwell, J. F. and Ellis, A. (2002) Transmission options for offshore wind farms in the United States, *Proceedings of the AWEC (American Wind Energy Association) Annual Conference*, 2002.

Wu, X., Arulampalam, A., Zhan, C. and Jenkins, N. (2003) Application of a static reactive power compensator (STATCOM) and a dynamic braking resistor (DBR) for the stability enhancement of a large wind farm, *Wind Engineering*, **27** (2), 93–106.

11

Wind Turbine Control for System Contingencies

11.1 Contribution of Wind Generation to Frequency Regulation

With the projected increase in wind generation, a potential concern for transmission system operators is the capability of wind farms to provide dynamic frequency support in the event of sudden changes in power network frequency (Eltra, 2004; E.ON Netz, 2006; National Grid, 2008).

11.1.1 Frequency Control

In any electrical power system, the active power generated and consumed has to be balanced in real time (on a second-by-second basis). Any disturbance to this balance causes a deviation of the system frequency. With an increase in demand, the system frequency will decrease and for a decrease in the demand the system frequency will increase. In many countries, the frequency delivered to consumers is maintained to within closer than $\pm 1\%$ of the declared value. For example, the system frequency of the England and Wales network under normal conditions is maintained at 50 Hz within operational limits of ± 0.2 Hz (Figure 11.1). This is achieved by operating the generators on a governor droop, normally around 4% and classified as the *continuous service* of the generator. However, if there is a sudden change in generation or load, the system frequency is allowed to deviate up to $+0.5$ Hz and -0.8 Hz.

In the event of a sudden failure in generation or connection of a large load, the system frequency starts to drop (region OX in Figure 11.1) at a rate mainly determined by the total angular momentum of the system (addition of

Wind Energy Generation: Modelling and Control   Olimpo Anaya-Lara, Nick Jenkins,
Janaka Ekanayake, Phill Cartwright and Mike Hughes
© 2009 John Wiley & Sons, Ltd

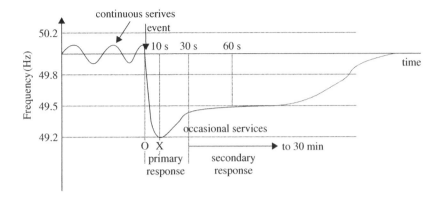

Figure 11.1   Frequency control in England and Wales (Erinmez *et al.*, 1999)

the angular momentum of all generators and spinning loads connected to the system). For the occasions when the frequency drops by more than 0.2 Hz, generation plants are contracted to provide additional frequency response. These response duties are classified as *occasional services* and have two parts, namely primary response and secondary response. In the UK, the primary and secondary response are defined as the additional active power that can be delivered from a generating unit that is available at 10 and 30 s, respectively, after an event and that can be sustained for 20 s to 30 min, respectively. Primary response is provided by an automatic droop control loop and generators increase their output depending on the dead band of their governor and time lag of their prime mover (e.g. that of the boiler drum in steam units). Secondary response is the restoration of the frequency back to its nominal value using a slow supplementary control loop. These services are illustrated in Figure 11.1.

11.1.2   Wind Turbine Inertia

An FSIG wind turbine acts in a similar manner to a synchronous machine when a sudden change in frequency occurs. For a drop in frequency, the machine starts to decelerate. This results in the conversion of kinetic energy of the machine to electrical energy, thus giving a power surge. The inverse is true for an increase in system frequency.

At any speed ω, the kinetic energy, E_k, in the rotating machine mass is given by the following equation:

$$E_k = \frac{1}{2} J \omega^2 \qquad (11.1)$$

If ω changes, then the power that is extracted is given by

$$P = \frac{dE_k}{dt} = \frac{1}{2}J \times 2\omega\frac{d\omega}{dt} = J\omega\frac{d\omega}{dt} \tag{11.2}$$

From Eq. (4.34), $J = 2S_{base}H/\omega_s^2$, hence from Eq. (11.2):

$$\frac{P}{S_{base}} = 2H\frac{\omega}{\omega_s}\frac{d(\omega/\omega_s)}{dt} \tag{11.3}$$

$$\overline{P} = 2H\overline{\omega}\frac{d\overline{\omega}}{dt} \tag{11.4}$$

Figure 11.2 shows the change in speed and electrical output power of a FSIG for a frequency step of 1 Hz. Commercial fixed-speed wind turbines rated above 1 MW have inertia constants, H, typically in the range of 3 to 5 s which illustrates the potential of wind turbines to contribute to fast frequency response.

In the case of a DFIG wind turbine, equipped with conventional controls, the control system operates to apply a restraining torque to the rotor according to a predetermined curve against rotor speed. This is decoupled from the power system frequency so there is no contribution to the system inertia. Figure 11.3 shows the change in speed and electrical output power for a frequency step of 1 Hz.

11.1.3 Fast Primary Response

With a large number of DFIG and/or FRC wind turbines connected to the network the angular momentum of the system will be reduced and so, the

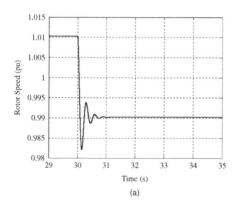
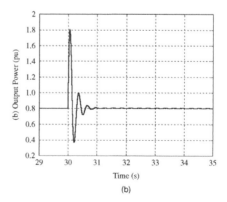

(a)   (b)

Figure 11.2 FSIG wind turbine. Change in output power for a step change in system frequency. (a) Rotor speed (pu); (b) Output power (pu)

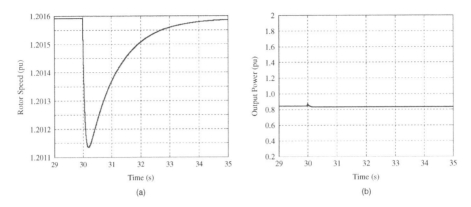

Figure 11.3 DFIG wind turbine with conventional controls. Change in output power for a step change in system frequency. (a) Rotor speed (pu); (b) Output power (pu)

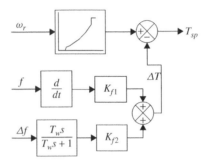

Figure 11.4 Supplementary control loop for machine inertia (Ramtharan *et al.*, 2007)

frequency may drop very rapidly during the period OX in Figure 11.1. There-fore, it is important to reinstate the effect of the machine inertia of these wind turbines. It is possible to emulate the inertia response by manipulating their control actions. The emulated inertia response provided by these generators is referred to as fast primary response.

As shown in Figure 11.4, two loops can be added to the DFIG or FRC current-mode controller to obtain fast primary response. One loop is pro-portional to the rate of change of frequency which represents the torque component given in Eq. (11.4) $(\overline{P}/\overline{\omega})$ and the other loop is fed through a washout term in proportion to a change in frequency.

Figure 11.5 shows the performance of the DFIG for a frequency drop with and without the second loop of Figure 11.4.

A block diagram of the FMAC plus auxiliary loop to obtain fast frequency response is shown in Figure 11.6. The operation of the auxiliary loop is as

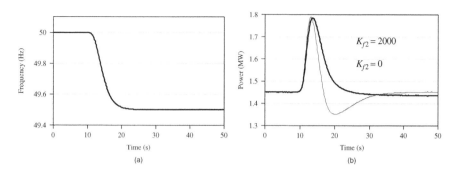

Figure 11.5 DFIG performance in response to a frequency event with $K_{f1} = 10\,000\,\text{Nm\,s}^2$ and K_{f2} as shown in Nms. (a) Power system frequency variation; (b) active power (Ramtharan *et al.*, 2007)

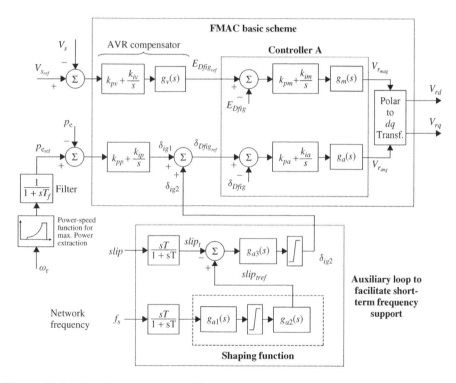

Figure 11.6 FMAC control plus auxiliary loop for frequency support (Anaya-Lara, *et al.*, 2006)

follows. When a loss of system generation occurs, the resulting fall in network frequency is measured and processed by the 'shaping function' block to generate a desired reference value, $slip_{tref}$, for the auxiliary slip control loop. The reference signal, $slip_{tref}$, defines the transient variation in slip that is desired to release rotor stored energy from the DFIG over the first few seconds following the loss of network generation. This differs from the approach adopted in the previous section with the current-mode controller, which employs a df_s/dt term. The reference signal, $slip_{tref}$, is obtained by passing the input network frequency signal initially through a washout element (to eliminate steady-state contribution) and then through a shaping element to provide the required transient profile for the slip reference set-point over the critical period whilst the frequency is low (Anaya-Lara et al., 2006).

The reference value is kept within limits to ensure that for the speed reduction demanded the turbine rotor is not driven into aerodynamic stall. The DFIG slip signal is also processed through a washout element. The error signal is processed through a simple lead–lag compensator to produce the output of the auxiliary loop, δ_{ig2}, which serves to increase the demanded value of $\delta_{Dfig_{ref}}$. The resulting increase in flux angle produces an increase in generator torque and a consequent reduction in rotor speed. The rotor speed is driven down to follow the transient swing in the reference value and is returned to the original value when the steady conditions are once again achieved.

The three-bus generic network model shown in Figure 8.12 was used to demonstrate the fast frequency support given by FMAC controller with and without the auxiliary loop and shown in Figure 11.7.

11.1.4 Slow Primary Response

To provide slow primary or secondary response from a generator, the generator power must increase or decrease with system frequency changes. Hence in order to respond to sustained low frequency, it is necessary to de-load the wind turbine leaving a margin for power increase.

11.1.4.1 Pitch Angle Control

Both FSIG and DFIG wind turbines can be de-loaded using a pitch angle power production control strategy. Using a conventional power production control strategy of pitch-to-feather the blade, the pitch angle is progressively reduced with the wind speed in order to maintain the rated output power (Ekanayake et al., 2003; Holdsworth et al., 2004a). Therefore, above rated power, if the pitch angle is controlled such that a fraction of the power that could be extracted from wind is 'spilled', this leaves a margin for additional loading of the wind turbine and hence the possibility to provide slow primary response.

Below rated power, the pitch of the blades is typically fixed at an optimum value, normally around $-2°$. However, in some variable-speed turbines the

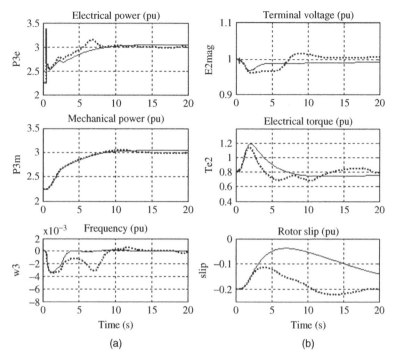

Figure 11.7 Frequency regulation. (a) Main system; (b) DFIG wind farm. Generator 2 is the DFIG with the FMAC basic scheme plus auxiliary loop. The DFIG is operating in super-synchronous mode with $s = -0.2$ pu (Anaya-Lara *et al.*, 2006)

pitch angle may be varied over a range of values for maximizing energy capture in light winds. This ensures that the rotor can extract the maximum available power from the prevailing wind speed. By changing the controlled-rotor pitch angle, it is possible to de-load the wind turbine. Figure 11.8 illustrates the effect of changing the pitch angle from $-2°$ to $+2°$ on the power extracted by the machine. It should be noted that, in comparison with the DFIG wind turbine, the changes in pitch angle affect the output power more dramatically on the FSIG wind turbine due to the rotor speed being fixed typically within 1% of synchronous.

For example, a FSIG wind turbine at a wind speed of $12 \, \text{m s}^{-1}$ operates at point X for a pitch angle of $+2°$ and at point Y for a pitch angle of $-2°$. In the case of a DFIG wind turbine, if the electronic controller operates so as to extract maximum power from the wind, the machine operates on the optimal power extraction line OA. For a wind speed of $12 \, \text{m s}^{-1}$ the DFIG wind turbine operates at point P for a pitch angle of $+2°$ and at point Q for a pitch angle of $-2°$.

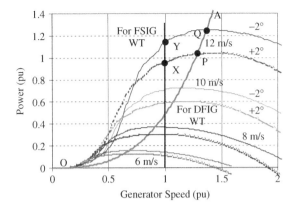

Figure 11.8 Effect of pitch angle control from $-2°$ to $+2°$ (Ekanayake *et al.*, 2003; Holdsworth *et al.*, 2004a)

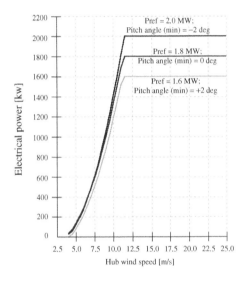

Figure 11.9 De-loading with pitch angle control (Holdsworth *et al.*, 2004a)

Figure 11.9 shows the effect of varying the minimum pitch angle for wind speeds below rated and the power production control reference point ($P_{reference}$) for wind speeds above rated speed. The figure shows that above rated wind speed, the power production controller reference power can be regulated to produce de-loading of 400 kW. Below rated wind speed, changing the pitch angle from $-2°$ to $+2°$ can offer de-loading of up to 400 kW. It should be noted that for wind speeds above rated, the pitch angle will be substantially larger than $+2°$.

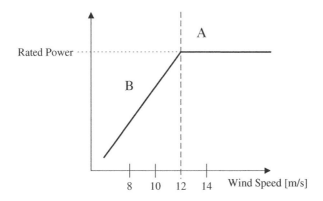

Figure 11.10 Operating regions for a pitch angle controller

The conventional power production controller for a pitch-regulated wind turbine operates in two regions, as shown in Figure 11.10. For both FSIG and DFIG wind turbines operating above rated output power (region A), the controller employs the pitch-to-feather power production control strategy to maintain the rated output power. For the FSIG wind turbine, the controller will operate below rated output power (region B) at a minimum pitch angle as specified by the control limits. For the DFIG wind turbine, the power electronic controller dominates the operation for region B. Below rated output power, the variable-speed wind turbine is controlled to operate on the predetermined torque–speed curve, which ensures maximum power extraction.

The same modified pitch angle controller for frequency response can be used for both FSIG and DFIG wind turbines. The control action can again be defined in the two regions as shown in Figure 11.10. In region A, the existing pitch-to-feather strategy is modified to implement the regulation of $P_{reference}$ required for low- and high-frequency response. In region B, the minimum pitch angle control limit is regulated for frequency response. However, for the DFIG wind turbine this control strategy operates in parallel with the power electronic controller. A block diagram of the pitch angle controller for frequency response is shown in Figure 11.11. Two droop controllers shown in Figure 11.11 are illustrated in Figure 11.12.

11.1.4.2 Electronic Control for DFIG Wind Turbines

If the rotor speed is changed so as to operate the machine off the optimal power extraction curve, then de-loading can be achieved using the electronic controller. Figure 11.13a shows how 10% de-loading can be achieved on the generator speed–power curve (for example, point Y to Z at $10\,\mathrm{m\,s^{-1}}$).

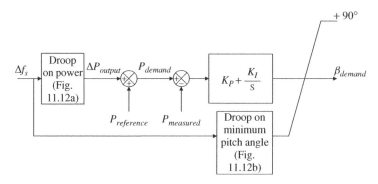

Figure 11.11 Pitch angle controller for frequency response from wind turbines (Holdsworth *et al.*, 2004a)

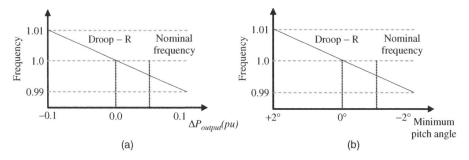

Figure 11.12 (a) Droop on power; (b) droop on minimum pitch angle (Holdsworth *et al.*, 2004a)

However, as the machine controller is based on the generator speed–torque curve, the same set of curves is transformed to the torque–speed plane as shown in Figure 11.13b (that is, from curve OAB to PQB) (Ekanayake *et al.*, 2003).

The torque–speed characteristic used for the DFIG wind turbine controller shown in Figure 5.8 (OAB in Figure 11.13b) can be replaced by the characteristic corresponding to 90% power extraction (PQB in Figure 11.13b) to obtain frequency response from the DFIG wind turbine. When the machine is de-loaded in this way, for a given wind speed the operating speed of the machine will be less than the speed corresponding to the 100% power case. The maximum de-loading possible when using the electronic control is approximately 90% of the rated power. For greater de-loading, pitch control is necessary.

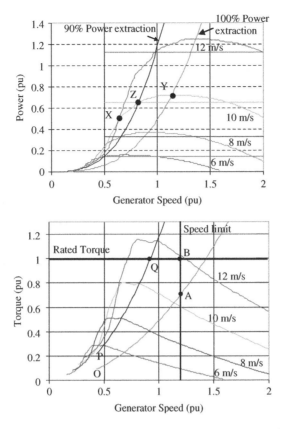

Figure 11.13 De-loading using electronic control. (a) 10% de-loading on power; (b) 10% de-loading on torque (Ekanayake *et al.*, 2003)

Figure 11.14 shows how the electronic controller may be used for frequency response. In this case, the machine should be de-loaded by modifying the set point torque curve (to PQB in Figure 11.13b) and then a droop on the electromagnetic torque can be added to the set-point torque to control the output of the machine.

The operation of this controller can be explained using Figure 11.13a. For example, assume that the wind turbine operates at point Z with de-loading to 90%. If the frequency drops, then the set-point torque initially decreases. As the mechanical torque on the shaft is constant, the wind turbine accelerates, hence the rotor speed increases towards point Y, giving 100% power output. On the other hand, if the frequency increases, then the reverse takes place, moving the operating point towards X.

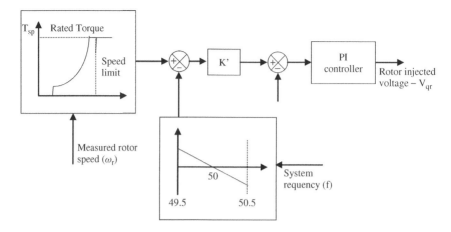

Figure 11.14   Frequency control of the DFIG (Ekanayake *et al.*, 2003)

11.2   Fault Ride-through (FRT)

11.2.1   FSIGs

A technique based on fast pitching of wind turbine blades, where the mechanical input torque is reduced throughout the duration of a power system disturbance, was presented by Holdsworth *et al.* (2004b) and Le and Islam (2008). A block diagram of the proposed 'fast-pitching' control strategy is shown in Figure 11.15. The 'fast pitching' is initiated by a fault flag.

The proposed 'fast-pitching' blade angle pitch controller operates in two modes:

- During normal operation, a standard power production control (PPC) strategy of pitch-to-feather is applied for power extraction (Burton *et al.*, 2001).

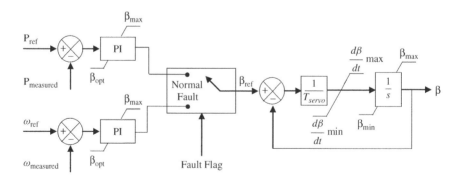

Figure 11.15   Fast-pitching scheme

- During abnormal operation, a speed control 'fast-pitching' strategy is applied. It increases the pitch angle subjected to a maximum rate, thus spilling the input mechanical power. Holdsworth *et al.* (2004b) proposed the rate of change of kinetic energy of the generator rotor as the fault flag.

11.2.2 DFIGs

The voltage at the terminals of a DFIG wind turbine drops significantly when a fault occurs in the power system, thus causing the electric power output of the generator to be greatly reduced. The mechanical input power is almost constant through the short duration of the fault and therefore the excess power (a) goes through the converters, thus increasing rotor currents, and (b) causes the machine to accelerate, thus storing kinetic energy in the rotating mass. The FRT capability of DFIG wind turbines may impose design challenges, especially in terms of:

- Protection of converters against over current and overvoltages. In order to protect the converters, a crowbar is normally employed. The crowbar triggers when the rotor current exceeds a threshold, thus short-circuiting the rotor of the DFIG.
- Minimizing the stresses on the mechanical shaft during the network disturbances.
- Minimizing or eliminating the reactive current absorption from the network during and after recovering from the fault.

One of the commonly employed mechanisms for fault ride-through (FRT) is based on a crowbar (Morren and Haan, 2005; Hansen and Michalke, 2007; Erlich *et al.*, 2007; Liu *et al.*, 2008). When the crowbar is triggered, the DFIG behaves as a fixed-speed induction generator with an increased rotor resistance. The insertion of the rotor resistance shifts the speed at which the pull-out torque occurs into higher speeds and reduces the amount of reactive power absorption. Both these aid the stability of the generator under a system fault.

Different types of crowbars are employed for a DFIG, namely:

1. *Soft crowbar*
   When the rotor fault current reaches the crowbar current limit, the crowbar, commonly referred to as a soft crowbar, short-circuits the rotor terminals through a high-energy dissipation resistor to reduce the rotor fault current and simultaneously opens the terminals of the rotor side converter. Once

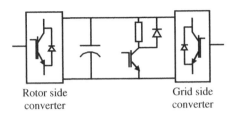

Rotor side   Grid side
converter   converter

Figure 11.16   Chopper arrangement

the rotor fault current has been reduced to an acceptable level, the crowbar by-passes the dissipating resistor and reconnects the rotor side converter to re-establish control over the generator. The soft crowbar performs this operation repeatedly for the duration of the fault.

2. *Single-shot crowbar*

   In this case, the crowbar performs only one operation during the fault, where it by-passes the rotor circuits to the dissipating resistor and keeps this state during the fault.

3. *Active crowbar*

   Instead of a passive crowbar, an active crowbar can be used to aid the DFIG wind turbine FRT. An active crowbar is essentially a controllable resistor controlled by an IGBT switch. A DC chopper as shown in Figure 11.16 can be used as an active crowbar. During a fault, when the rotor current exceeds a certain limit, the IGBTs will be blocked. However, the current continues to flow into the DC link through the freewheeling diodes leading to an increase in the DC link voltage. To keep the DC link voltage below the upper threshold, the chopper is switched ON.

11.2.2.1   DFIG FRT Performance with Soft Crowbar

Figure 11.17 shows a snapshot of the terminal voltage and rotor current during the fault where the operation of the soft crowbar is seen in detail. The crowbar current limit is set to $I_{r\,max} = 3.0\,pu$. The on and off operation of the soft crowbar can be observed as a chattering in the voltage and torque responses during the fault.

A snapshot of the terminal voltage and rotor current during the fault with single-shot crowbar is shown in Figure 11.18. When the crowbar current limit is reached, the crowbar by-passes the rotor circuit through the dissipating resistor and as a result the rotor fault current decreases gradually during the fault. When the single-shot crowbar is used, the DFIG may become unstable if the fault is sustained for a longer time due to the lack of control.

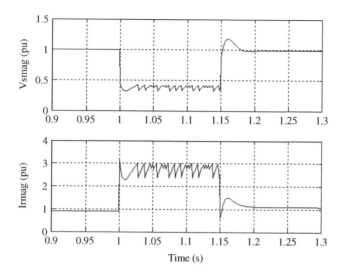

Figure 11.17 DFIG responses for a fault applied at $t = 1$ s with a fault clearance time of 150 ms. Soft crowbar in operation. $I_{r\,max} = 3.0$ pu; $V_{s\,mag}$ 0.32–0.42 pu; $I_{r\,mag}$ 2.4–3.0 pu

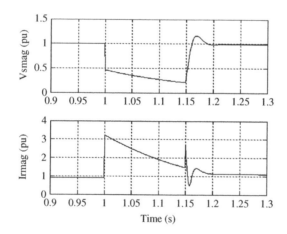

Figure 11.18 DFIG responses for a fault applied at $t = 1$ s with a fault clearance time of 150 ms. Single-shot crowbar in operation. $I_{r\,max} = 3.0$ pu

11.2.3   FRCs

In order to assess the dynamic performance of the network-side converter in the event of faults, the test system illustrated in Figure 11.19 is used. Here, the network-side converter controls the active and reactive power flows to the grid where an inductive reactance X_s is connected between the converter and

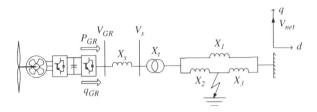

Figure 11.19 Network to assess the performance of the network-side converter in the event of faults

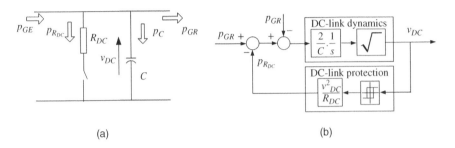

Figure 11.20 (a) Power flow in the DC link; (b) DC link dynamic representation

the turbine terminals. The turbine is then connected to a transformer and then to the network through a double line circuit.

   In the event of a network fault, the DC link voltage rises rapidly because the network-side converter cannot transfer all the active power coming from the generator. Therefore, a chopper resistor protection system is used to dissipate the excessive energy in the DC link (Figure 11.20) (Conroy and Watson, 2007; Ramtharan, 2008). The DC link is short-circuited through the resistor R_{DC} when the DC link voltage exceeds the maximum limit. The voltage across the capacitor is determined by considering the power balance at the DC link and given by:

$$v_{DC}^2 = \frac{2}{C} \int (p_{GE} - p_{R_{DC}} - p_{GR})\mathrm{d}t \qquad (11.5)$$

Therefore, the DC link dynamics are represented as shown in Figure 11.20b.

   Another approach for FRT of an FRC wind turbine is de-loading the generator. This will reduce the power transferred from the wind turbine to the DC link. Therefore, the excess power is stored as kinetic energy in the rotating mass, thus increasing the wind turbine speed. The FRC wind turbine is de-loaded by multiplying the torque reference by a quantity which is proportional to the DC capacitor voltage as shown in Figure 11.21. When the DC

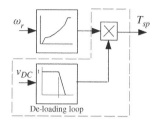

Figure 11.21 De-loading droop for wind turbine fault ride-through (Ramtharan, 2008)

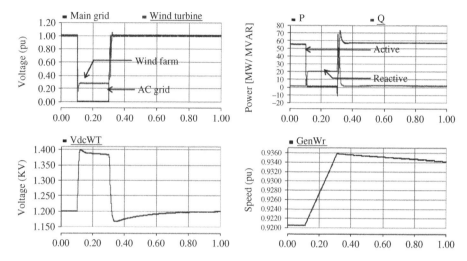

Figure 11.22 FRT performance with de-loaded WT (Ramtharan, 2008). (a) Wind turbine and main grid voltage; (b) wind turbine active and reactive power; (c) DC capacitor voltage; (d) generator speed

capacitor voltage is above a threshold, the reference torque is multiplied by a fractional value, thus reducing the power extracted by the generator-side converter.

Figure 11.22 shows the performance of the controller shown in Figure 11.21 for a three-phase short-circuit fault of 200 ms duration.

11.2.4   *VSC–HVDC with FSIG Wind Farm*

Figure 11.23 shows an FSIG wind farm connected through a VSC–HVDC link. In normal operation, the offshore converter is controlled to maintain the voltage and frequency of the wind farm network at its nominal values. The onshore converter is controlled to regulate the DC link voltage. This ensures that the energy collected from the offshore converter is transmitted to the

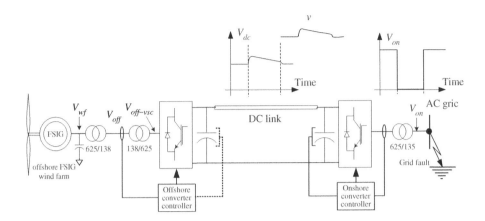

Figure 11.23 FSIG with VSC–HVDC connected to grid

onshore AC network. The onshore converter also provides reactive power support to the grid.

During a terrestrial system fault, the onshore converter is unable to deliver active power to the grid. The DC link voltage increases rapidly and this would cause the HVDC link to trip on overvoltage. Therefore, real power from the wind farm has to be reduced to control the DC link voltage. This can be achieved by increasing the rotor speed, which allows increasing wind energy to be stored as kinetic energy in the rotating inertia of the wind turbine rotors during a fault. For a fixed-speed induction machine, the rotational speed can be increased by reducing the electromagnetic torque. In the steady state, the electromagnetic torque of a FSIG is given by Eq. (4.6). From this equation, it is clear that the electromagnetic torque can be reduced either by reducing the terminal voltage or by increasing the synchronous speed by raising the frequency of the offshore network (Xu *et al.*, 2007; Arulampalam *et al.*, 2008).

11.2.5 FRC Wind Turbines Connected Via a VSC–HVDC

An FRC wind farm connected to the AC system through a VSC–HVDC as shown in Figure 11.24 was considered. The network-side HVDC converter maintains the HVDC link voltage close to the specified reference level by adjusting the active power transmitted to the AC network to match that received from the wind farm-side HVDC converter. The wind farm-side HVDC converter maintains the voltage and frequency of the wind farm AC network.

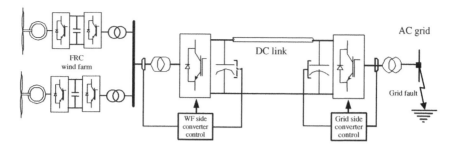

Figure 11.24 Fully rated converter wind farm connected through a VSC HVDC

During a fault in the AC network, the HVDC link voltage increases rapidly. Therefore, in order to maintain the HVDC link voltage below its upper limit, the wind farm output power has to be reduced. The following three methods are considered for reducing the wind farm power output.

11.2.5.1 Rapid De-loading a Fully Rated Converter Wind Turbine

FRC wind turbines can be de-loaded in two different ways to facilitate HVDC fault ride-through. One way is to reduce the generator torques via generator-side converter control. An alternative is to block the output power via the wind turbines' HVDC-side converter control through setting the active power current components to zero.

De-loading Via the Generator Controller
Figure 11.25 shows how the HVDC de-loading droop is incorporated into the wind turbine de-loading droop already shown in Figure 11.21. The scheme

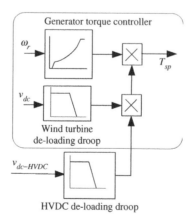

Figure 11.25 De-loading a fully rated converter wind turbine via generator controller

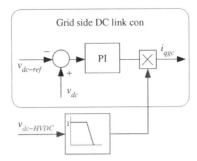

Figure 11.26 De-loading a fully rated converter wind turbine via HVDC converter side controller

ensures that the torque set point is directly reduced when the HVDC link voltage increases beyond its threshold value.

De-loading Via Wind Turbine HVDC Converter Side Controller

In Figure 11.26, the HVDC de-loading droop introduces a multiplying factor into the wind turbine HVDC converter-side active power current controller. When the HVDC link voltage increases beyond its threshold value, the wind turbines' HVDC converter-side active power current is reduced to block the wind turbine output power. Reducing the power at the wind turbine HVDC converter-side converter increases the wind turbine DC link voltage and in turn activates the wind turbine de-loading controller.

De-loading the FRC wind turbines via the wind turbine generator controllers or via the wind turbine network-side controllers ultimately increases the rotor speed and converts the aerodynamic power into kinetic energy.

It was assumed that an ideal communication medium is available between the offshore HVDC converter and every wind turbine to dispatch the de-loading signals. Modern wind farms use fast communication circuits such as fibre optics for the SCADA system; however, the availability of such communication links for fault ride-through remains a question. Therefore, in practice there may be some delay in sending the de-loading signals (due to the unavailability of communication channel) to each wind turbine.

11.2.5.2   Emulated Short-circuiting of the Wind Farm Side HVDC Converter

The difficulties in relying on dedicated control signals from HVDC link to de-load individual turbines when an AC fault occurs can be avoided by adopting an alternative approach. The location of the AC fault can be effectively

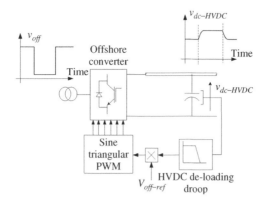

Figure 11.27   Short-circuiting controller in the offshore HVDC converter

transferred to the wind farm side by reducing the voltage at the wind farm side-HVDC converter terminal.

When an AC fault occurs, the increase in HVDC link voltage is detected to adjust the amplitude modulation index of the wind farm-side HVDC converter. Figure 11.27 shows the emulated short-circuit protection controller incorporated into the sinusoidal PWM of the wind farm side HVDC converter. The de-loading droop gain, acting on the HVDC link voltage, is multiplied by the reference wind farm-side HVDC converter voltage. As the HDVC link voltage increases during a network-side fault, the de-loading controller reduces the amplitude modulation index and then the terminal voltage of the wind farm-side HVDC converter is effectively reduced. Therefore, the wind farm sees an apparent fault in the wind farm network when there is, in fact, a fault in the main AC network.

11.2.5.3   Chopper Resistor on the HVDC Link

An alternative approach to maintaining the HVDC link voltage below the upper limit during an AC network fault is to dissipate the excess power as heat. A chopper resistor may be used on the HVDC link to dissipate the wind farm input power during an AC network fault.

References

Anaya-Lara, O., Hughes, F. M., Jenkins, N. and Strbac, G. (2006) Contribution of DFIG-based wind farms to power system short-term frequency regulation, *IEE GTD Proceedings*, **153** (2), 164–170.

Arulampalam, A., Ramtharan, G., Caliao, N., Ekanayake, J. B. and Jenkins, N. (2008) Simulated onshore-fault ride through offshore wind farms connected through VSC HVDC, *Wind Engineering*, **32** (2), 103–113.

Burton, T., Sharpe, D., Jenkins, N. and Bossanyi, E. (2001) *Wind Energy Handbook*, John Wiley & Sons, Ltd, Chichester, ISBN 0 471 48997 2.

Conroy, J. F. and Watson, R. (2007) Low-voltage ride-through of a full converter wind turbine with permanent magnet generator, *IET RPG Proceedings*, **1** (3), 182–189.

Ekanayake, J. B., Holdsworth, L. and Jenkins, N. (2003) Control of doubly fed induction generator (DFIG) wind turbine, *IEE Power Engineering*, **17** (1), 28–32.

Eltra (2004) *Wind Turbines Connected to Grids with Voltages Above 100 kV*, Technical Regulations TF 3.2.5, Doc. No. 214493 v3, Eltra, Skærbæk.

E.ON Netz (2006) *Grid Connection Regulations for High and Extra High Voltage*, E.ON Netz GmbH, Bayreuth.

Erinmez, I.A., Bickers, D.O., Wood, G.F. and Hung, W. W. (1999) NGC experience with frequency control in England and Wales – provision of frequency response by generator, presented at the IEEE PES Winter Meeting.

Erlich, I., Wilch, M. and Feltes, C. (2007) Reactive power generation by dfig based wind farms with AC grid connection, presented at EPE 2007 – 12th European Conference on Power Electronics and Applications, 2–5 September 2007, Aalborg, Denmark.

Hansen, A. D. and Michalke, G. (2007) *Fault ride-through capability of DFIG wind turbines*, *Renewable* Energy, **32**, 1594–1610.

Holdsworth, L., Ekanayake, J. B. and Jenkins, N. (2004a) Power system frequency response from fixed speed and doubly fed induction generator-based wind turbines, *Wind Energy*, **7** (1), 21–35.

Holdsworth, L., Charalambous, I., Ekanayake, J. B. and Jenkins, N. (2004b) Power system fault ride through capabilities of induction generator based wind turbines, *Wind Engineering*, **28** (4), 399–409.

Le, H. N. D. and Islam, S. (2008) Substantial control strategies of DFIG wind power system during grid transient faults, *IEEE/PES Transmission and Distribution Conference and* Exposition, 2008, T&D, pp. 1–13.

Liu, Z., Anaya-Lara, O., Quinonez-Varela G. and McDonald, J. R. (2008) Optimal DFIG crowbar resistor design under different controllers during grid faults, *Third International Conference on Electric Utility Deregulation and Restructuring and Power Technologies*, DRPT 2008.

Morren, J. and de Haan, S. W. H. (2005) Ride through of wind turbines with doubly-fed induction generator during a voltage dip, *IEEE Transactions on Energy Conversion*, **20** (2), 435–441.

National Grid (2008) *The Grid Code*, Issue 3, Revision 25.

Ramtharan, G. (2008) Control of variable speed wind turbine generators. PhD Thesis. University of Manchester.

Ramtharan, G., Ekanayake, J. B. and Jenkins, N. (2007) Frequency support from doubly fed induction generator wind turbines, *IET Renewable Power Generation*, **1** (1), 3–9.

Xu, L., Yao, L. and Sasse, C. (2007) Grid integration of large DFIG-based wind farms using VSC transmission, *IEEE Transactions on Power Systems*, **22** (3), 976–984.

Appendix A

State–Space Concepts and Models

State–space concepts in dynamic systems can readily be demonstrated in terms of a simple example. Consider the simple electrical network shown in Figure A.1, which consists of an inductor, L, a resistor, R, a capacitor, C, and a voltage source, e.

In terms of current i, the network equation can be expressed as

$$L\frac{di}{dt} + Ri + \frac{1}{C}\int i\,dt = e$$

or in terms of capacitor charge, q, since $i = dq/dt$:

$$L\frac{d^2q}{dt^2} + R\frac{dq}{dt} + \frac{1}{C}q = e$$

This second-order differential equation can be solved analytically to determine i or q as a function of time. For a complex system, however, a high-order differential equation would be obtained, for which an analytical solution would not be feasible. It is therefore, desirable to convert a high-order differential equation into a standard and more convenient form to facilitate analysis and solution.

This can be achieved by making use of Moigno's auxiliary variable method, which enables a single nth-order differential equation to be converted into n first-order differential equations.

In terms of the circuit example concerned (Figure A.1), we can set

$$q = x_1 \text{ and } \frac{dq}{dt} = x_2$$

Wind Energy Generation: Modelling and Control   Olimpo Anaya-Lara, Nick Jenkins,
Janaka Ekanayake, Phill Cartwright and Mike Hughes
© 2009 John Wiley & Sons, Ltd

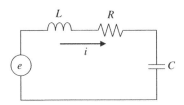

Figure A.1   Simple electrical *LRC* network

giving

$$L\frac{\mathrm{d}x_2}{\mathrm{d}t} + Rx_2 + \frac{1}{C}x_1 = e$$

and

$$\frac{\mathrm{d}x_1}{\mathrm{d}t} = x_2$$

Separating the derivative terms on the left-hand side and defining $u = e$ enables these equations to be written in the standard first-order form as follows:

$$\frac{\mathrm{d}x_1}{\mathrm{d}t} = x_2$$

$$\frac{\mathrm{d}x_2}{\mathrm{d}t} = -\frac{1}{LC}x_1 - \frac{R}{L}x_2 + \frac{1}{L}u$$

If x_1 and x_2 are known, then the precise condition or 'state' of the system is known. Consequently, x_1 and x_2 are referred to as the system 'state variables' and the equations themselves as the system 'state equations'.

A dynamic system is one which consists totally or, in part, of energy storage elements, and since energy cannot change instantaneously it is these elements which cause the system variables to be time dependent and require the system to be modelled in terms of differential equations. Also, the order n of a dynamic system, that is, the number of first-order differential equations required to represent it, is determined by the number of independent energy storage elements in the system. This fact can readily be demonstrated by deriving the equations of the simple network (Figure A.1) from energy concepts instead of network theory.

In the circuit in Figure A.1, we have two energy storage elements, the inductor and the capacitor. The circuit stored energy, E, is given by

$$E = \frac{1}{2}Li^2 + \frac{1}{2}\frac{1}{C}q^2$$

Of the other two circuit elements, the voltage source injects power, whereas the resistor dissipates power. The circuit power P is given by

$$P = ei - Ri^2$$

Power is also given by the rate of change of energy:

$$\frac{dE}{dt} = Li\frac{di}{dt} + \frac{1}{C}q\frac{dq}{dt}$$

Since

$$P = \frac{dE}{dt}$$

$$ei - Ri^2 = Li\frac{di}{dt} + \frac{1}{C}q\frac{dq}{dt}$$

Noting that $dq/dt = i$ and dividing throughout by i leads to the network equation:

$$e = L\frac{di}{dt} + Ri + \frac{1}{C}q$$

In general, each independent energy storage element will have associated with it an independent system variable. Consequently, from a power balance approach to equation derivation, on differentiating stored energy, that is, forming dE/dt, the number of differentiated system variables will be given by the number of independent energy storage elements. The number of first-order differential equations in the resulting model, therefore, will also be determined by the number of independent energy storage elements.

The number of state variables associated with a state–space model of a dynamic system is given by the number of independent energy storage elements in the system.

Vectors

Before defining what is meant by a state vector let us consider the general definition of a vector with the aid of Figure A.2.

A vector can, therefore, be defined as an ordered set of numbers. This puts no limit on dimensionality. In general:

$$\mathbf{x} = (x_1, x_2, \ldots, x_n) \text{ or } \mathbf{x} = \begin{bmatrix} x_1 \\ x_2 \\ \vdots \\ x_n \end{bmatrix}$$

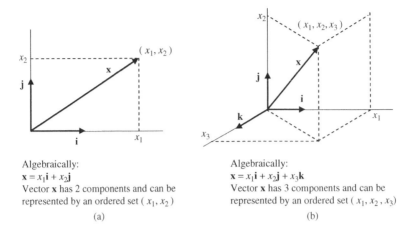

Algebraically:
$\mathbf{x} = x_1\mathbf{i} + x_2\mathbf{j}$
Vector **x** has 2 components and can be
represented by an ordered set (x_1, x_2)

(a)

Algebraically:
$\mathbf{x} = x_1\mathbf{i} + x_2\mathbf{j} + x_3\mathbf{k}$
Vector **x** has 3 components and can be
represented by an ordered set (x_1, x_2, x_3)

(b)

Figure A.2   Definition of a vector. (a) Two dimensions; (b) three dimensions

where **x** is an n-dimensional vector in n-dimensional space and can be defined
in terms of a row or column vector.

Matrix Form of State Equations

The previously defined state equations, since they are linear, can be manipulated into the very convenient and standard matrix form, as shown below.
   The relationship between the system inputs and states can be written as

$$\begin{bmatrix} \dot{x}_1 \\ \dot{x}_2 \end{bmatrix} = \begin{bmatrix} 0 & 1 \\ -\dfrac{1}{LC} & -\dfrac{R}{L} \end{bmatrix} \begin{bmatrix} x_1 \\ x_2 \end{bmatrix} + \begin{bmatrix} 0 \\ \dfrac{1}{L} \end{bmatrix} [u]$$

and the relationship between the system states and the output can be written as

$$[y] = \begin{bmatrix} \dfrac{1}{C} & 0 \end{bmatrix} \begin{bmatrix} x_1 \\ x_2 \end{bmatrix}$$

The above equations have the general form

$$\dot{\mathbf{x}} = \mathbf{A}\mathbf{x} + \mathbf{B}\mathbf{u}$$

$$\mathbf{y} = \mathbf{C}\mathbf{x}$$

where
$\mathbf{x} = [x_1, x_2, \ldots, x_n]^T$ is the state vector of order n;

$\mathbf{u} = [u_1, u_2, \ldots, u_r]^T$ is the input vector of order r;

$\mathbf{y} = [y_1, y_2, \ldots, y_m]^T$ is the output vector of order m;

and

  A is an $n \times n$ state matrix;

  B is an $n \times r$ input matrix;

  C is an $m \times n$ output matrix.

Matrix Operations

Active

With this type of operation (Figure A.3), vectors are changed into new vectors, for example $\mathbf{y} = \mathbf{Bx}$; matrix **B** changes vector **x** into vector **y**.

$$\text{Let } \mathbf{B} = \begin{bmatrix} 0 & 1 \\ -4 & 2 \end{bmatrix}$$

and if

$$\mathbf{x} = \begin{bmatrix} x_1 \\ x_2 \end{bmatrix} = \begin{bmatrix} 1 \\ 1 \end{bmatrix}$$

then

$$\mathbf{y} = \mathbf{Bx} = \begin{bmatrix} 0 & 1 \\ -4 & 2 \end{bmatrix}\begin{bmatrix} 1 \\ 1 \end{bmatrix} = \begin{bmatrix} 1 \\ -2 \end{bmatrix}$$

  In terms of a dynamic system model, $\dot{\mathbf{x}} = \mathbf{Ax}$; matrix **A** changes vector **x** into vector $\dot{\mathbf{x}}$ (Figure A.4).

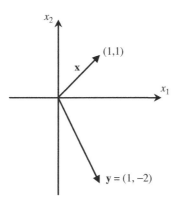

Figure A.3   Vector representation

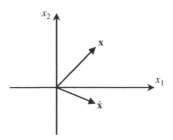

Figure A.4   Matrix changes on a vector

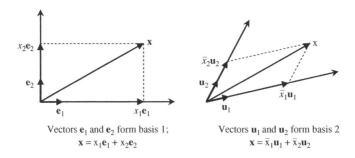

Vectors $\mathbf{e}_1$ and $\mathbf{e}_2$ form basis 1;   Vectors $\mathbf{u}_1$ and $\mathbf{u}_2$ form basis 2
$$\mathbf{x} = x_1\mathbf{e}_1 + x_2\mathbf{e}_2$$ $$\mathbf{x} = \bar{x}_1\mathbf{u}_1 + \bar{x}_2\mathbf{u}_2$$

Figure A.5   Same vector expressed in two different ways

Passive

In this case, a matrix merely changes the description of some object from an 'old' description to a 'new' description.

Passive operators are used for changes of basis (Figure A.5). Both bases map out the same two-dimensional state space involved.

The relationship between coordinates in basis 1 to coordinates in basis 2 is given by a *passive operator* as follows:

$$\begin{bmatrix} \bar{\mathbf{x}}_1 \\ \bar{\mathbf{x}}_2 \end{bmatrix} = [\mathbf{T}] \begin{bmatrix} x_1 \\ x_2 \end{bmatrix}$$

Basis

A basis of a vector space is any coordinate set which generates the space (Figure A.6).

Consider now how the coordinates of $\mathbf{x}$ with respect to the basis frame defined by axes $\mathbf{e}_1$ and $\mathbf{e}_2$ are related to the coordinates of the same vector

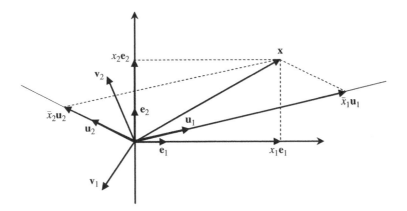

Figure A.6 Basis of a vector space

when expressed with respect to the basis frame defined by axes $\mathbf{u}_1$ and $\mathbf{u}_2$:

$$\mathbf{x} = x_1\mathbf{e}_1 + x_2\mathbf{e}_2 \text{ and } \mathbf{x} = \bar{x}_1\mathbf{u}_1 + \bar{x}_2\mathbf{u}_2$$

Let us find out how new coordinates $\bar{x}_1$ and $\bar{x}_2$ are related to x_1 and x_2. Introduce a reciprocal set of vectors $\mathbf{v}_1$ and $\mathbf{v}_2$ such that

$$\mathbf{v}_1^t\mathbf{u}_1 = 1 \text{ and } \mathbf{v}_1^t\mathbf{u}_2 = 0$$
$$\mathbf{v}_2^t\mathbf{u}_1 = 0 \text{ and } \mathbf{v}_2^t\mathbf{u}_2 = 1$$

that is, vector $\mathbf{v}_1$ is orthogonal to $\mathbf{u}_2$ and vector $\mathbf{v}_2$ is orthogonal to $\mathbf{u}_1$:

$$\begin{bmatrix} \mathbf{v}_1^t & \cdots \\ \mathbf{v}_2^t & \cdots \end{bmatrix}\begin{bmatrix} \mathbf{u}_1 & \mathbf{u}_2 \\ \vdots & \vdots \end{bmatrix} = \begin{bmatrix} 1 & 0 \\ 0 & 1 \end{bmatrix}$$

$$\mathbf{VU} = \mathbf{I}; \quad \mathbf{V} = \mathbf{U}^{-1}$$

New basis frame axes $\mathbf{u}_1$ and $\mathbf{u}_2$ can be expressed in terms of old basis axes $\mathbf{e}_1$ and $\mathbf{e}_2$:

$$\mathbf{u}_1 = u_{11}\mathbf{e}_1 + u_{12}\mathbf{e}_2; \quad \mathbf{u}_1 = \begin{bmatrix} u_{11} \\ u_{12} \end{bmatrix}$$

$$\mathbf{u}_2 = u_{21}\mathbf{e}_1 + u_{22}\mathbf{e}_2; \quad \mathbf{u}_2 = \begin{bmatrix} u_{21} \\ u_{22} \end{bmatrix}$$

Hence

$$
\begin{bmatrix} \mathbf{v}_1^t & \cdots \\ \mathbf{v}_2^t & \cdots \end{bmatrix} = \begin{bmatrix} \mathbf{u}_1 & \mathbf{u}_2 \\ \vdots & \vdots \end{bmatrix}^{-1} = \begin{bmatrix} u_{11} & u_{21} \\ u_{12} & u_{22} \end{bmatrix}^{-1} = \begin{bmatrix} v_{11} & v_{12} \\ v_{21} & v_{22} \end{bmatrix}
$$

that is,

$$
\mathbf{v}_1^t = [v_{11} \quad v_{12}]
$$

$$
\mathbf{v}_1 = \begin{bmatrix} v_{11} \\ v_{12} \end{bmatrix}
$$

To obtain new coordinates $\mathbf{x}_1$ and $\mathbf{x}_2$:

$$
\mathbf{v}_1^t \mathbf{x} = \mathbf{v}_1^t [\overline{x}_1 \mathbf{u}_1 + \overline{x}_2 \mathbf{u}_2] = \overline{x}_1
$$

$$
\mathbf{v}_2^t \mathbf{x} = \mathbf{v}_2^t [\overline{x}_1 \mathbf{u}_1 + \overline{x}_2 \mathbf{u}_2] = \overline{x}_2
$$

that is,

$$
\begin{bmatrix} \overline{x}_1 \\ \overline{x}_2 \end{bmatrix} = \begin{bmatrix} \mathbf{v}_1^t & \cdots \\ \mathbf{v}_2^t & \cdots \end{bmatrix} \begin{bmatrix} x_1 \\ x_2 \end{bmatrix}
$$

$$
\overline{\mathbf{x}} = \mathbf{V}\mathbf{x}
$$

Matrix $\mathbf{V}$ is a passive operator.

Appendix B

Introduction to Eigenvalues and Eigenvectors

Consider the dynamic system

$$\dot{x} = Ax$$

where

$$A = \begin{bmatrix} 0 & 1 \\ -2 & -3 \end{bmatrix}$$

Compute $dx = Ax dt$ as vector x traverses a unit circle (Figure B.1).

It is found that for two directions of vector x, vector dx points directly at the origin, that is, x and dx are in the same direction. The effect of operator A is simply to make dx a *scalar multiple* of x. In the special case

$$A\mu = \lambda\mu$$

λ is the scalar multiplier (eigenvalue) and μ is the special vector (eigenvector).

Calculation of $\dot{x} = Ax$ as x moves around the unit circle:

1. $x^t = [0 \quad 1]$;
$$\dot{x} = \begin{bmatrix} 0 & 1 \\ -2 & -3 \end{bmatrix}\begin{bmatrix} 0 \\ 1 \end{bmatrix} = \begin{bmatrix} 1 \\ -3 \end{bmatrix}$$

2. $x^t = \dfrac{1}{\sqrt{2}}[1 \quad 1]$;
$$\begin{bmatrix} \dot{x}_1 \\ \dot{x}_2 \end{bmatrix} = \begin{bmatrix} 0 & 1 \\ -2 & -3 \end{bmatrix}\begin{bmatrix} 1 \\ -1 \end{bmatrix}\frac{1}{\sqrt{2}} = \frac{1}{\sqrt{2}}\begin{bmatrix} 1 \\ -5 \end{bmatrix}$$

3. $x^t = [1 \quad 0]$;
$$\begin{bmatrix} \dot{x}_1 \\ \dot{x}_2 \end{bmatrix} = \begin{bmatrix} 0 & 1 \\ -2 & -3 \end{bmatrix}\begin{bmatrix} 1 \\ 0 \end{bmatrix} = \begin{bmatrix} 0 \\ -2 \end{bmatrix}$$

Wind Energy Generation: Modelling and Control   Olimpo Anaya-Lara, Nick Jenkins, Janaka Ekanayake, Phill Cartwright and Mike Hughes
© 2009 John Wiley & Sons, Ltd

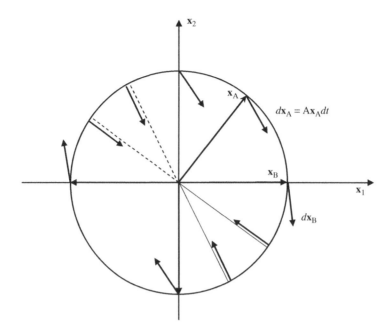

Figure B.1   Unit circle

4. $\mathbf{x}^t = \begin{bmatrix} 1 & -1 \end{bmatrix}\dfrac{1}{\sqrt{2}}$;

$$\begin{bmatrix} \dot{x}_1 \\ \dot{x}_2 \end{bmatrix} = \begin{bmatrix} 0 & 1 \\ -2 & -3 \end{bmatrix}\begin{bmatrix} 1 \\ -1 \end{bmatrix}\dfrac{1}{\sqrt{2}} = \begin{bmatrix} -1 \\ +1 \end{bmatrix}\dfrac{1}{\sqrt{2}}$$

5. $\mathbf{x}^t = \begin{bmatrix} 1 & -2 \end{bmatrix}\dfrac{1}{\sqrt{5}}$;

$$\begin{bmatrix} \dot{x}_1 \\ \dot{x}_2 \end{bmatrix} = \begin{bmatrix} 0 & 1 \\ -2 & -3 \end{bmatrix}\begin{bmatrix} 1 \\ -2 \end{bmatrix}\dfrac{1}{\sqrt{5}} = \begin{bmatrix} -2 \\ +4 \end{bmatrix}\dfrac{1}{\sqrt{5}}$$

6. $\mathbf{x}^t = \begin{bmatrix} 0 & -1 \end{bmatrix}$;

$$\begin{bmatrix} \dot{x}_1 \\ \dot{x}_2 \end{bmatrix} = \begin{bmatrix} 0 & 1 \\ -2 & -3 \end{bmatrix}\begin{bmatrix} 0 \\ -1 \end{bmatrix} = \begin{bmatrix} -1 \\ +3 \end{bmatrix}$$

Eigenvalues and Eigenvectors

The dynamic properties of our simple system $\dot{\mathbf{x}} = \mathbf{A}\mathbf{x}$ can be much more readily observed and analysed if instead of using the original basis we choose as basis the eigenvectors of the system.

Let

$$\bar{\mathbf{x}} = \mathbf{V}\mathbf{x}$$

that is,

$$\mathbf{x} = \mathbf{V}^{-1}\bar{\mathbf{x}} = \mathbf{U}\bar{\mathbf{x}}$$

Now,

$$\dot{\mathbf{x}} = \mathbf{Ax}$$
$$\dot{\overline{\mathbf{x}}} = \mathbf{V}\dot{\mathbf{x}} = \mathbf{VAx} = \mathbf{VAU}\overline{\mathbf{x}}$$

Consider the matrix **VAU**. If matrix **U** is comprised of columns of eigenvectors, that is,

$$\mathbf{U} = \begin{bmatrix} \mathbf{u}_1 & \mathbf{u}_2 \\ \vdots & \vdots \end{bmatrix} = \begin{bmatrix} u_{11} & u_{21} \\ u_{12} & u_{22} \end{bmatrix}$$

Then, as demonstrated for each eigenvector.

$$\mathbf{Au_i} = \lambda_i \mathbf{u_i}$$

Therefore,

$$\mathbf{VAU} = \begin{bmatrix} \mathbf{v}_1^t & \cdots \\ \mathbf{v}_2^t & \cdots \end{bmatrix} [\mathbf{A}] \begin{bmatrix} \mathbf{u}_1 & \mathbf{u}_2 \\ \vdots & \vdots \end{bmatrix} = \begin{bmatrix} \mathbf{v}_1^t & \cdots \\ \mathbf{v}_2^t & \cdots \end{bmatrix} \begin{bmatrix} \lambda_1\mathbf{u}_1 & \lambda_2\mathbf{u}_2 \\ \vdots & \vdots \end{bmatrix}$$

$$= \begin{bmatrix} \lambda_1 & 0 \\ 0 & \lambda_2 \end{bmatrix} = \Lambda \quad \text{since } \mathbf{v}_1^t\mathbf{u}_1 = 1, \ \mathbf{v}_1^t\mathbf{u}_2 = 0, \text{ etc.}$$

Hence,

$$\begin{bmatrix} \dot{\overline{x}}_1 \\ \dot{\overline{x}}_2 \end{bmatrix} = \begin{bmatrix} \lambda_1 & 0 \\ 0 & \lambda_2 \end{bmatrix} \begin{bmatrix} \overline{x}_1 \\ \overline{x}_2 \end{bmatrix}$$

Stability

When a model of a dynamic system is expressed with respect to an eigenvector basis frame, its state matrix is diagonal in form:

$$\begin{bmatrix} \dot{\overline{x}}_1 \\ \dot{\overline{x}}_2 \end{bmatrix} = \begin{bmatrix} \lambda_1 & 0 \\ 0 & \lambda_2 \end{bmatrix} \begin{bmatrix} \overline{x}_1 \\ \overline{x}_2 \end{bmatrix}$$

Consequently, the state equations are independent of each other, that is,

$$\frac{d\overline{x}_1}{dt} = \overline{x}_1 \text{ and } \frac{d\overline{x}_2}{dt} = \overline{x}_2$$

The time response solutions, from initial conditions $\overline{x}_1(0)$ and $\overline{x}_2(0)$, are therefore

$$\overline{x}_1(t) = \overline{x}_1(0)e^{\lambda_1 t}$$
$$\overline{x}_2(t) = \overline{x}_2(0)e^{\lambda_2 t}$$

It can be seen that the stability of a linear system is determined by its eigen-values.

If the real part of any eigenvalue λ_i is positive, then the associated mode $\overline{x}_i(t)$ will approach infinite magnitude as $t \to \infty$.

Calculation of Eigenvalues and Eigenvectors

The system equation is

$$\dot{\mathbf{x}} = \mathbf{A}\mathbf{x}$$

If $\mathbf{u_i}$ are system eigenvectors and λ_i are system eigenvalues, then

$$\mathbf{A}\mathbf{u_i} = \lambda_i \mathbf{u_i}$$

so that

$$\mathbf{A}\mathbf{u_i} - \lambda_i \mathbf{u_i} = 0$$

Hence $[\mathbf{A} - \lambda_i \mathbf{I}]\mathbf{u_i} = 0$
The above can only be true (for non-zero $\mathbf{u_i}$) if

$$\det [\mathbf{A} - \lambda_i \mathbf{I}] = 0$$

(so that $[\mathbf{A} - \lambda_i \mathbf{I}]^{-1}$ does not exist).

The system eigenvalues are determined by solving the above polynomial equation in λ.

Consider the simple system where

$$\mathbf{A} = \begin{bmatrix} 0 & 1 \\ -2 & -3 \end{bmatrix}$$

Then

$$\mathbf{A} - \lambda\mathbf{I} = \begin{bmatrix} 0 & 1 \\ -2 & -3 \end{bmatrix} - \begin{bmatrix} \lambda & 0 \\ 0 & \lambda \end{bmatrix} = \begin{bmatrix} -\lambda & 1 \\ -2 & (-3-\lambda) \end{bmatrix}$$

$$\det [\mathbf{A} - \lambda\mathbf{I}] = \lambda(\lambda + 3) + 2 = \lambda^2 + 3\lambda + 2 = (\lambda + 1)(\lambda + 2)$$

Eigenvalues are given as solutions when $\det[\mathbf{A} - \lambda\mathbf{I}] = 0$, that is, $\lambda_1 = -1$ and $\lambda_2 = -2$, matrix $\mathbf{A}$ in fact is representative of a simple LRC network (Figure B.2).

$$\begin{bmatrix} \dot{x}_1 \\ \dot{x}_2 \end{bmatrix} = \begin{bmatrix} 0 & 1 \\ -\dfrac{1}{LC} & -\dfrac{R}{L} \end{bmatrix}; \qquad \dfrac{1}{LC} = 2, \qquad \dfrac{R}{L} = 3$$

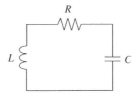

Figure B.2 Simple *LRC* network

Calculation of Eigenvectors

These are calculated from $\mathbf{Au_i} = \lambda \mathbf{u_i}$

First, consider $\lambda_1 = -1$:

$$\mathbf{Au_i} = \lambda_i \mathbf{u_i} \text{ or } (\mathbf{A} - \lambda_i \mathbf{I})\mathbf{u_i} = 0$$

$$\left\{ \begin{bmatrix} 0 & 1 \\ -2 & -3 \end{bmatrix} - (-1) \begin{bmatrix} 1 & 0 \\ 0 & 1 \end{bmatrix} \right\} \begin{bmatrix} u_{11} \\ u_{12} \end{bmatrix} = \begin{bmatrix} 1 & 1 \\ -2 & -2 \end{bmatrix} \begin{bmatrix} u_{11} \\ u_{12} \end{bmatrix} = \begin{bmatrix} 0 \\ 0 \end{bmatrix}$$

that is

$$u_{11} + u_{12} = 0; \quad u_{11} = -u_{12}$$

Let $u_{11} = 1$, then $u_{12} = -1$. Hence

$$\mathbf{u_i} = \begin{bmatrix} 1 \\ -1 \end{bmatrix}$$

Second, consider another eigenvalue, $\lambda_2 = -2$:

$$\left\{ \begin{bmatrix} 0 & 1 \\ -2 & -3 \end{bmatrix} - (-2) \begin{bmatrix} 1 & 0 \\ 0 & 1 \end{bmatrix} \right\} \begin{bmatrix} u_{21} \\ u_{22} \end{bmatrix} = \begin{bmatrix} 2 & 1 \\ -2 & -1 \end{bmatrix} \begin{bmatrix} u_{21} \\ u_{22} \end{bmatrix}$$

$$= \begin{bmatrix} 0 \\ 0 \end{bmatrix} 2u_{21} + u_{22} = 0$$

that is,

$$\mathbf{u_2} = \begin{bmatrix} 1 \\ -2 \end{bmatrix}$$

NB. If $\mathbf{u}$ is an eigenvector, then so is $(\alpha \mathbf{u})$ for any α.

Now,

$$\mathbf{U} = \begin{bmatrix} \mathbf{u_1} & \mathbf{u_2} \\ \vdots & \vdots \end{bmatrix} = \begin{bmatrix} 1 & 1 \\ -1 & -2 \end{bmatrix} : \text{matrix of eigenvectors}$$

$$V = U^{-1} = \begin{bmatrix} 2 & 1 \\ -1 & -1 \end{bmatrix} = \begin{bmatrix} v_1^t & \cdots \\ v_2^t & \cdots \end{bmatrix}$$

$$v_1 = \begin{bmatrix} 2 \\ 1 \end{bmatrix}; \quad v_2 = \begin{bmatrix} -1 \\ -1 \end{bmatrix}$$

As a check on the solution, if

$$\Lambda = VAU$$

then

$$V^{-1}\Lambda U^{-1} = V^{-1}VAUU^{-1} = A$$

Form:

$$U\Lambda V = \begin{bmatrix} 1 & 1 \\ -1 & -2 \end{bmatrix} \begin{bmatrix} -1 & 0 \\ 0 & -2 \end{bmatrix} \begin{bmatrix} 2 & 1 \\ -1 & -1 \end{bmatrix}$$

$$= \begin{bmatrix} 0 & 1 \\ -2 & -3 \end{bmatrix} = A$$

Appendix C

Linearization of State Equations

In general, state equations are of the form

$$\frac{dx_i}{dt} = f_i(x_1, x_2, \ldots, x_n) \quad i = 1, 2, \ldots, n$$

At a singular point (or equilibrium point):

$$\frac{dx}{dt} = 0$$

If the singular point is $x^t = c^t = [c_1, c_2, \ldots, c_n]$, then

$$f_i(c_1, c_2, \ldots, c_n) = 0 \text{ for } i = 1, 2, \ldots, n$$

The behaviour of the system in the vicinity of the singular point is required (Figure C.1).

The state equations can now be re-expressed in terms of the small deviations about the singular point to give

$$\frac{dx_i}{dt} = \frac{d}{dt}(c_i + y_i) = f_i(c_1 + y_1, c_2 + y_2, \ldots, c_n + y_n)$$

These can be expanded in terms of a Taylor series to give

$$\frac{d}{dt}(c_i + y_i) = f_i(c_1, c_2, \ldots, c_n) + \sum_{r=1}^{n} \frac{\partial f_i}{\partial x_r} y_r + \sum_{r=1}^{n} \sum_{s=1}^{n} \frac{\partial^2 f_i}{\partial x_r \partial x_s} y_r y_s + \cdots$$

Now, $f_i(c_1, c_2, \ldots, c_n = 0)$ by definition of the singular point. Hence, by choosing the deviation vector y to be sufficiently small, the second-order term

Wind Energy Generation: Modelling and Control   Olimpo Anaya-Lara, Nick Jenkins,
Janaka Ekanayake, Phill Cartwright and Mike Hughes
© 2009 John Wiley & Sons, Ltd

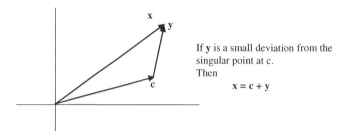

If $\mathbf{y}$ is a small deviation from the
singular point at c.
Then

$$\mathbf{x} = \mathbf{c} + \mathbf{y}$$

Figure C.1   Behaviour in the vicinity of a singular point

and higher order terms become insignificant in the Taylor expansion. The state
equations can then be approximated as

$$\frac{\mathrm{d}}{\mathrm{d}t}(x_i) = \frac{\mathrm{d}}{\mathrm{d}t}(y_i) \approx \sum_{r=1}^{n} \frac{\partial f_i}{\partial x_r} y_r \text{ for } i = 1, 2, \ldots, n$$

and a set of linear state equations has been achieved.

Example

Consider the nonlinear system having the state equations

$$\frac{\mathrm{d}x_1}{\mathrm{d}t} = -x_1 + x_2{}^4 = f_1$$

$$\frac{\mathrm{d}x_2}{\mathrm{d}t} = 2x_1 - 3x_2 + x_1^4 = f_2$$

A singular point exists at the origin, since when $x_1 = 0, x_2 = 0$ both $\mathrm{d}x_1/\mathrm{d}t$
and $\mathrm{d}x_2/\mathrm{d}t$ are zero.

Prior to assessing stability in the vicinity of the origin, the nonlinear
equations need to be linearized:

$$\frac{\partial f_1}{\partial x_1} = -1; \quad \frac{\partial f_1}{\partial x_2} = 4x_2{}^3$$

$$\frac{\partial f_2}{\partial x_1} = 2 - 4x_1{}^3; \quad \frac{\partial f_2}{\partial x_2} = -3$$

At the singular point $x_1 = x_2 = 0$, hence

$$\frac{\partial f_1}{\partial x_1} = -1; \quad \frac{\partial f_1}{\partial x_2} = 0$$

$$\frac{\partial f_2}{\partial x_1} = 2; \quad \frac{\partial f_2}{\partial x_2} = -3$$

giving the linearized model in the region of the singular point as

$$\begin{bmatrix} \dot{y}_1 \\ \dot{y}_2 \end{bmatrix} = \begin{bmatrix} -1 & 0 \\ 2 & -3 \end{bmatrix} \begin{bmatrix} y_1 \\ y_2 \end{bmatrix} \rightarrow \dot{\mathbf{y}} = \mathbf{A}\mathbf{y}$$

Stability can be determined from the eigenvalues of these linearized equations. If any eigenvalues have positive real parts, then the system is unstable. and if all eigenvalues have negative real parts, then it is stable.

$$\det(\lambda\mathbf{I} - \mathbf{A}) = 0 = \begin{Vmatrix} \lambda + 1 & 0 \\ -2 & \lambda + 3 \end{Vmatrix} = 0$$

$$\det = (\lambda + 1)(\lambda + 3) = 0$$

hence

$$\lambda_1 = -1, \quad \lambda_2 = -3$$

Both roots have negative real parts, therefore the system is stable in the vicinity of the singular point at the origin.

Appendix D

Generic Network Model Parameters

In the responses included in Chapters 8 and 9, the generator, excitation systems, turbine and governor systems are identical.

Parameters (on base of machine rating) of Generator 1 and Generator 3

$$X_d = 2.13,\ X'_d = 0.308,\ X''_d = 0.234$$
$$X_q = 2.07,\ X'_q = 0.906,\ X''_q = 0.234,\ X_P = 0.17$$
$$T'_{d0} = 6.0857\,\text{s},\ T''_{d0} = 0.0526\,\text{s},\ T'_{q0} = 1.653\,\text{s}$$
$$T''_{q0} = 0.3538\,\text{s},\ H = 3.84\,\text{s}$$

Excitation Control System (Generator 1 and Generator 3)

This is shown in Figure D.1.

Steam Turbine and Governor Parameters (Generator 1 and Generator 3)

These are shown in Figure D.2.

Wind Energy Generation: Modelling and Control   Olimpo Anaya-Lara, Nick Jenkins,
Janaka Ekanayake, Phill Cartwright and Mike Hughes
© 2009 John Wiley & Sons, Ltd

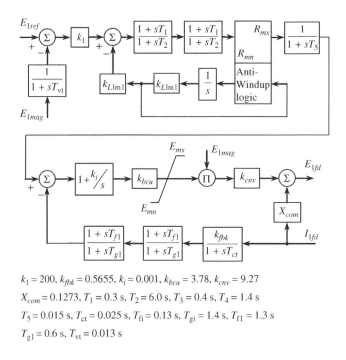

$k_1 = 200$, $k_{fbk} = 0.5655$, $k_i = 0.001$, $k_{bcu} = 3.78$, $k_{cnv} = 9.27$

$X_{com} = 0.1273$, $T_1 = 0.3$ s, $T_2 = 6.0$ s, $T_3 = 0.4$ s, $T_4 = 1.4$ s

$T_5 = 0.015$ s, $T_{ct} = 0.025$ s, $T_{fi} = 0.13$ s, $T_{gi} = 1.4$ s, $T_{f1} = 1.3$ s

$T_{g1} = 0.6$ s, $T_{vt} = 0.013$ s

Figure D.1   Excitation control system

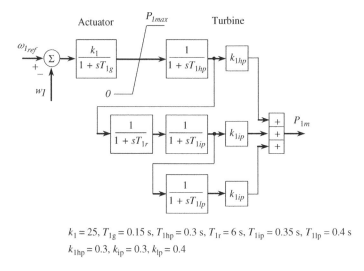

$k_1 = 25$, $T_{1g} = 0.15$ s, $T_{1hp} = 0.3$ s, $T_{1r} = 6$ s, $T_{1ip} = 0.35$ s, $T_{1lp} = 0.4$ s

$k_{1hp} = 0.3$, $k_{ip} = 0.3$, $k_{lp} = 0.4$

Figure D.2   Steam turbine and governor parameters

FSIG and DFIG Parameters (on Base of Machine Rating)

$$R_s = 0.00488, \ R_r = 0.00549, \ X_{1s} = 0.09241$$

$$X_{1r} = 0.09955, \ X_{1m} = 3.95279, \ H = 3.5 \, \text{s}$$

Generic Network Parameters

$X11 = 0.14$ (on base of generator 1 rating)

$X21 = 0.137$ (on base of generator 2 rating)

$X12 = 0.01, \ X22 = 0.1337, \ X3 = 0.2$ (on base Sb = 1000 MVA)

DFIG Parameters (on Base of Machine Rating, $S_b = 700 \, \text{MVA}$) Chapter 5

$R_s = 0.00488, \ R_r = 0.00549, \ X_{1s} = 0.09241$

$X_{1r} = 0.09955, \ X_{1m} = 3.95279, \ H = 3.5 \, \text{s}$

Converter power rating 25% of DFIG

Nominal slip range ±20%

Control Parameters and Transfer Functions for the FMAC Controller

Voltage loop: $k_{pv} = 5; \ k_{iv} = 0.5; \ g_v(s) = \dfrac{1 + 0.024s}{1 + 0.004s} \cdot \dfrac{1 + 0.035s}{1 + 0.05s}$

Power loop: $k_{pp} = 0.4; \ k_{ip} = 0.05; \ T_f = 1 \, \text{s}$

Wash-out time constant $T = 5 \, \text{s}$

Controller A: $g_m(s) = g_a(s) = \left(\dfrac{1 + 0.4s}{1 + 2s}\right); \ k_{pm} = k_{pa} = 1.2; \ k_{im} = k_{ia} = 0.01$

Generator Parameters Used for the Simulation Chapter 6

Rating: 2 MVA

Line-to-line voltage: 966 V

Poles: 64

Rated frequency: 50 Hz

Total moment of inertia of generator and turbine: $7.5 \times 10^3 \, \text{kgm}^2$

Parameters: $r_s = 0.00234 \, \Omega, \quad X_{1s} = 0.1478 \, \Omega, \quad X_q = 0.6017 \, \Omega, \quad X_d = 0.6017 \, \Omega, \quad r_{fd} = 0.0005 \, \Omega, \quad X_{1fd} = 0.2523 \, \Omega, \quad r_{kq2} = 0.01675 \, \Omega, \quad X_{1kq2} = 0.1267 \, \Omega, \ r_{kd} = 0.01736 \, \Omega, \ X_{1kd} = 0.1970 \, \Omega, \ r_{kq1}$ and X_{1kq1} not present

DC Link Parameters

Reference DC link voltage V_{DC-ref}: 1000 V
Capacitor C: 10 mF

FSIG Turbine Parameters Chapter 7

Rotor diameter: 35 m
Number of blades: 3
Blade inertia about shaft: 42 604 kg m^2
Hub inertia about shaft: 1500 kg m^2
Total rotor inertia: 129 312 kg m^2
Gearbox ratio: 32:1
Squirrel cage induction generator:
  Ratings: 300 kW
  Generator inertia: 100 kg m^2
  Line-to-line voltage: 415 V
  Poles: 4
  Rated frequency: 50 Hz
  Parameters: $r_s = 0.004\,\Omega$, $r_r = 0.0032\,\Omega$, $X_{ls} = 0.0383\,\Omega$, $X_{lr} = 0.0772\,\Omega$, $X_m = 1.56\,\Omega$

DFIG Turbine Parameters

Rotor diameter: 75 m
Number of blades: 3
Blade inertia about shaft: 1.419×10^6 kg m^2
Hub inertia about shaft: 12 000 kg m^2
Total rotor inertia: 4.268×10^6 kg m^2
Gearbox ratio: 84.15:1
Doubly fed induction generator:
  Ratings: 2 MW
  Generator inertia: 130 kg m^2
  Lint-to-line voltage: 690 V

Poles: 4

Rated frequency: 50 Hz

Parameters: $r_s = 0.001164\,\Omega$, $r_r = 0.00131\,\Omega$, $X_{ls} = 0.022\,\Omega$, $X_{lr} = 0.0237\,\Omega$, $X_m = 0.941\,\Omega$

Index

Wind Energy Generation: Modelling and Control   Olimpo Anaya-Lara, Nick Jenkins,
Janaka Ekanayake, Phill Cartwright and Mike Hughes
© 2009 John Wiley & Sons, Ltd

Printed and bound by CPI Group (UK) Ltd, Croydon, CR0 4YY

27/10/2024

14580307-0003